Even More Brain-Powered Science

Teaching and Learning With Discrepant Events

Even More Brain-Powered Science

Teaching and Learning With Discrepant Events

Thomas O'Brien

NSTApress

National Science Teachers Association

Arlington, VA

National Science Teachers Association

Claire Reinburg, Director
Jennifer Horak, Managing Editor
Andrew Cooke, Senior Editor
Judy Cusick, Senior Editor
Wendy Rubin, Associate Editor
Amy America, Book Acquisitions Coordinator

ART AND DESIGN
Will Thomas Jr., Director, Cover and interior design
Cover and inside illustrations by Daniel Vasconcellos unless noted otherwise

PRINTING AND PRODUCTION
Catherine Lorrain, Director

NATIONAL SCIENCE TEACHERS ASSOCIATION
Francis Q. Eberle, PhD, Executive Director
David Beacom, Publisher

LIBRARY OF CONGRESS CATALOGING-IN-PUBLICATION DATA
O'Brien, Thomas.
  Even more brain-powered science : teaching and learning with discrepant events / by Thomas O'Brien.
     p. cm.
  Includes bibliographical references and index.
  ISBN 978-1-936137-21-3
  1. Science--Study and teaching (Elementary)--Activity programs. 2. Science--Study and teaching
(Secondary)--Activity programs. I. Title.
  LB1585.O287 2011
  372.35'044--dc22
                        2011000444

e-ISBN 978-1-936137-50-3

NSTA is committed to publishing material that promotes the best in inquiry-based science education. However, conditions of actual use may vary, and the safety procedures and practices described in this book are intended to serve only as a guide. Additional precautionary measures may be required. NSTA and the authors do not warrant or represent that the procedures and practices in this book meet any safety code or standard of federal, state, or local regulations. NSTA and the authors disclaim any liability for personal injury or damage to property arising out of or relating to the use of this book, including any of the recommendations, instructions, or materials contained therein.

PERMISSIONS
Book purchasers may photocopy, print, or e-mail up to five copies of an NSTA book chapter for personal use only; this does not include display or promotional use. Elementary, middle, and high school teachers may reproduce forms, sample documents, and single NSTA book chapters needed for classroom or noncommercial, professional-development use only. E-book buyers may download files to multiple personal devices but are prohibited from posting the files to third-party servers or websites, or from passing files to non-buyers. For additional permission to photocopy or use material electronically from this NSTA Press book, please contact the Copyright Clearance Center (CCC) (*www.copyright.com*; 978-750-8400). Please access *www.nsta.org/permissions* for further information about NSTA's rights and permissions policies.

About the Cover—Safety Issues: In the cartoon drawing on the cover, artist Dan Vasconcellos depicts the energy and excitement present in a school science lab. During actual school lab investigations, students should always maintain a safe distance from the teacher who is doing the demonstration. The teacher and students should wear personal protection equipment if the demonstration has any potential for bodily harm. Safety Notes throughout this book spell out when a demonstration requires that the teacher and students wear safety goggles or other protective items.

Contents

Contents

Acknowledgments

As I complete this third book in my *Brain-Powered Science* series, I once again owe a debt of gratitude to my former science teachers, past and present colleagues, students, and authors who have informed and inspired my passion for teaching and learning with discrepant events.

I would also like to especially thank members of the NSTA Press staff with whom I've had the pleasure to work with and learn from since my initial meeting with Claire Reinburg, director of NSTA Press, at the NSTA National Conference in Boston in March 2008. Claire's interest in my dual-purpose use of discrepant-event activities as model science-inquiry lessons and visual participatory analogies for science teacher education encouraged me to continue refining my preliminary draft. Rather than restrict the scope of my vision, she wisely counseled me to think in terms of a book series rather than a single book. Without her encouragement and support, this series would not be in print. NSTA editors Judy Cusick (*Brain-Powered Science*) and Wendy Rubin (*More Brain-Powered Science* and *Even More Brain-Powered Science*) and illustrator Daniel Vasconcellos deserve credit for their contributions to the book's written and visual appeal. Finally, I appreciate the ongoing efforts of the NSTA marketing department to promote this book to middle and high school science teachers and college science teacher educators.

Finally, I would like to acknowledge the unnamed reviewers and teacher-readers of this book for your dedication to your students' learning and to your own professional development and that of your peers. Your transformational leadership inside and outside your classroom, department, and school can play a catalytic role in making world-class science curriculum-instruction-assessment the norm in U.S. schools. Thank you for investing your money and time in buying, reading, using, and sharing this book. With your further refinement, I believe these activities will generate rich dividends for you, your students, and your colleagues.

About the Author

Dr. Thomas O'Brien's 33 years in science education began in K–12 schools, where he taught general, environmental, and physical sciences and high school chemistry. For the past 23 years, he has directed the preservice and inservice graduate-level science teacher–education programs of the School of Education at Binghamton University (State University of New York [SUNY]). His master's-level courses include Philosophical and Theoretical Foundations of Science Teaching, Curriculum and Teaching in Science, and Elementary Science Content and Methods. He also supervises the student teaching practica. In addition, he teaches a cross-listed doctoral/post-master's educational leadership course.

Concurrent with and subsequent to earning a master's degree and doctorate in Curriculum and Instruction/Science Education at the University of Maryland–College Park, Dr. O'Brien served as a curriculum development specialist and Teacher's Guide editor on the first edition of the American Chemical Society's *Chemistry in the Community* (1988) textbook and as the coauthor of the *New York Science, Technology & Society Education Project Teacher Guide* (1996).

As a science teacher professional development specialist, he has co-taught more than 25 summer institutes, including national programs of the Institute for Chemical Education and state and regional programs funded by grants from the National Science Foundation, the Howard Hughes Medical Institute, and the New York State Education Department, among others. He has received awards for excellence in teaching and/or service from the American Chemical Society (for National Chemistry Week programs), the New York State Association of Teacher Educators, the SUNY chancellor, and the New York State Science Education Leadership Association. These grants and awards are a reflection of collaborations with university-based colleagues and what he has learned with and from the large number of K–12 teachers he has had the privilege to serve. The *Brain-Powered Science* book series owes a debt of gratitude to these friends and funding agencies for the insights and opportunities they offered the author.

Introduction

This is the third book in a series designed for grades 5–12 preservice and inservice science teachers and teacher educators. Like its predecessors, this volume features *discrepant-event* activities that can be used for two related but distinct purposes. First, these initially anomalous demonstration-experiments are designed as models of interactive, inquiry-oriented activities (NRC 2000; NSTA 2004a) for teachers to experience as "learners" *and* use as teachers in their own classrooms. Engaging learners' attention and natural curiosity and catalyzing a need-to-know motivation are the first steps to unleashing the power of the human mind. As such, these activities elicit unanswered questions and challenge unquestioned answers about the content and nature of science. Second, these same discrepant-event activities are also presented as *visual participatory analogies for science teacher education*. Used as analogies, they create an experiential context and conceptual bridge for teachers to critically examine principles of research-informed, standards-based, best-practice curriculum-instruction-assessment (CIA).

The synergy created by combining these two effective instructional strategies opens the doors to both enhanced teacher and student learning (see Appendix A for other minds-on strategies for teaching [MOST]). More than being simply a compilation of engaging science "tricks," this book series supports thought-provoking, work-embedded professional development that emphasizes action in and critical reflection on practices that make a difference in the teacher-readers' classrooms. To help ensure this applicability, the activities are designed to be safe, simple, economical, enjoyable, effective, and relevant for use in both teacher professional development *and* grades 5–12 classroom contexts (O'Brien 2010, Appendix A).

As science teachers, we work in an era of American and world history that could certainly be considered both the best of times and the worst of times. Interrelated economic, energy, environmental, educational, and ethical challenges are cited in both the daily news and scholarly reports. These challenges can be viewed either as crises that require an attribution of blame or as opportunities to apply research-informed, system-level strategies for changing the way we do things. Or, as Albert Einstein supposedly said, "The significant problems we face cannot be solved at the same level of thinking we were at when we created them" (Calaprice 2011, p. 476). In the education domain, numerous reports document that more intelligent

Introduction

science CIA systems are needed to meet the challenges of the 21st century (Appendix B; Hilton 2010; NAS 2007; NCEE 1983; NCMST 2000; NSB 2006; Partnership for 21st Century Skills 2009; Rotherham and Willingham 2010; STEM Education Coalition).

Research-informed pedagogical models that synergistically integrate versus simplistically sequence CIA (e.g., the 5E Teaching Cycle) and innovative education technologies are making "science for all Americans" an achievable goal rather than simply an idealistic slogan. "Mile-wide, inch-deep" curricula that "leave students with fragmented elements of knowledge and little sense of the intellectual and creative achievements of science or its explanatory coherence … [and] understanding of the practices of science and engineering" (NRC 2010a) are recognized as being in need of replacement. Preliminary work on grade-level and cross-grade-level *learning progressions* and a renewed focus on what are now being called *cross-cutting science concepts* promise to inform and reform science textbooks and state-mandated tests (Appendix C; NRC 2007, 2010a). Improvements in textbooks, technologies, and tests notwithstanding, research identifies a fourth *T*, the teacher, as the most critical external element for improving student achievement (NCMST 2000; NCTAF 1996, 1997; NRC 2010b; NSB 2006; NSTA 2007a; PCAST 2010). Caring, competent, and committed teachers inform, instruct, and inspire student-learners. The motivational and cognitive scaffolding teachers offer to students helps them co-construct the foundation for a more promising future. The cement that holds this foundation together is teachers' professional development, which extends from preservice education to a mentored induction period for novice teachers and career-long inservice education.

Unfortunately, "[t]here is a mismatch between the kind of teaching and learning teachers are now expected to pursue with their students and the teaching they experience in their own professional education" (NCTAF 1996, p. 84). "Professional development that supports student learning is rooted in the science that teachers teach and includes opportunities to learn about science, about current research on how children learn science and about how to teach science" (NRC 2007, p. 296). This book series' dual-purpose science and science-education activities provide a resource for creating these kinds of learning opportunities for teachers. Whether you are a preservice or an inservice teacher seeking a stimulus for self-reflection or a catalyst for collegial collaboration, the *Brain-Powered Science* series is for you. The embedded, holographic nature of the science education concepts allows the individual activities to be used in any sequence within and across the series. The books also provide an experiential base and models from which teachers can learn how to adapt and improve (versus merely adopt and use) science-inquiry activities found on the internet and in other science demonstration books (e.g., Becker 1993; Bell 2008; Bilash 1997; Bilash and Shields 2001; Bilash, Gross, and Koob 2006; Camp and Clement 1994; Cunningham and Herr 2002; Ehrlich 1990, 1997; Freier and Anderson 1981; Gross, Holzer, and Colangelo 2001; Ingram 2003; Jewett 1994, 1996; Kardos 2003; Keeley, Eberle, and Farrin 2005; Keeley, Eberle, and Tugel 2007; Sae 1996; Sarquis, Williams, and Sarquis 1995; Sarquis et al. 2009, 2010; Stepans 1994; Summerlin and Ealy 1988; Taylor, Poth, and Portman 1995; VanCleave 1989, 1990, 1991). Teacher-readers are also encouraged to expand their use of science analogies (Gilbert and Watt Ireton 2003; Hackney and Wandersee 2002; Harrison and Coll 2008; Lawson 1993).

Contexts in Which to Use This Book Series (For New Readers)

Preservice Science Teaching Methods Courses

The science-education focus of this three-book series correlates with topics that are typically covered in middle and secondary science-education methods courses. Any combination of the books can be used as either activity-oriented *supplements* to popular science teaching methods books or, if combined with instructor-selected readings, as a *substitute* for such books (e.g., Bybee, Carlson Powell, and Trowbridge 2008; Chiappetta and Koballa 2010; Gallagher 2007; Lawson 2010). College science teacher educators can use the dual-purpose activities with their preservice students as "walk the talk" models of inquiry-oriented, discrepant event–centered science teaching and/or as *visual participatory analogies* that provide a common experiential base to catalyze conversations about cognitive learning theory and research-informed best practices. The preservice students can use the activities in their roles as teachers-in-training during in-class microteach presentations and during their student teaching. In the latter context, section III's focus on 5E mini units is especially helpful in that it challenges the misconception that great teaching is about individual lessons rather than well-designed, multiday instructional sequences (and ultimately a course's overall scope and sequence). Similarly, section II's focus on reading in science challenges the misconception that equates "hands-off" with "minds-off."

Introduction

Grades 5–12 Science Classrooms: A Place for Teacher Learning

Effective science teaching depends on a teacher's ability to construct multi-lane, bidirectional bridges that connect the subject matter and students. Teachers must develop, creatively apply, and continually nurture their own science pedagogical content knowledge (PCK). PCK includes

> the most useful forms of representation of those [discipline specific] ideas, the most powerful analogies, illustrations, examples, explanations, and demonstrations—in a word, the ways of representing and formulating the subject that make it comprehensible to others … [It] also includes an understanding of what makes the learning of specific concepts easy or difficult: the conceptions and preconceptions that students of different ages and backgrounds bring with them to the learning. (Shulman 1986, p. 9; see also Cochran 1997; Hagevvik et al. 2010; NSTA 2003, 2004b; Shulman 1987)

Unfortunately, especially during the first few years of their careers, when they need the most support, many novice inservice teachers have limited time and external support for discipline-specific professional development. Additionally, more than one third of new teachers leave the profession within their first five years (Ingersoll 2003). Though this high attrition rate supports calls for high-quality, formalized mentoring and induction programs (Rhoton and Bowers 2003), individual teachers also can learn from their own daily planning and practice on the job.

This book is not written for casual, "on your seat" reading. Instead, the activities should be implemented on your feet in grades 5–12 classrooms and refined based on the teacher-readers' reflections on their own performance, their students' reactions, and the validity of the underlying pedagogical theories and strategies. The Science Education Concepts and Debriefing components of each activity encourage critical reflective practice (Schön 1983) that has direct transfer implications for teachers' CIA practices. The Internet Connections and Extensions provide support for each teacher's own self-directed professional development (as well as for collegial collaborations).

Professional Development Learning Communities for Inservice Teachers

An assumption of schools is that an individual student's cognitive growth (i.e., learning) is best supported in socially interactive, experientially rich, collaborative environments (i.e., classrooms) with scaffolded support provided by a master teacher (i.e., leader). Similarly, if schools aspire to be learning organizations for their students, they also need to be learning organizations for their teaching staff by deliberate design rather than happenstance. Research on building professional learning communities (PLCs) within and across schools and districts has been occurring for more than a decade (Banilower et al. 2006; DuFour and Eaker 1998; Garet et al. 2001; Hord and Summers 2008; Loucks-Horsley et al. 1998; Mundry and Stiles 2009; NRC 2001a; NSDC 2001; NSTA 2006, 2007a, 2007b; Tobias and Baffert 2010; Yager 2005). Different PLC models vary in the degree to which they provide centralized leadership from internal curriculum or professional development specialists and external consultants. However, most PLCs include a focus on curriculum-instruction-assessment and some form of lesson study as a kind of collegial research and development where teachers give and receive critical friends-type feedback on their actual practice (Coalition of Essential Schools Northwest; Dubin 2009; Lewis and Tsuchida 1998; Stigler and Hiebert 1999). Teacher-reflective practice and professional collaboration are also featured in national standards for teaching (e.g., Standards #9 and #10 in CCSSO 2010; Standards X and XI in NBPTS 2003a; Standards XII and XIII in NBPTS 2003b).

This book series supports both small and informal or larger, more formal teachers-helping-teachers networks that encourage dynamic two-way interactions between theory and practice within and between middle and secondary schools as applied research and development learning laboratories, and colleges and universities that offer preservice and inservice teacher education. The dual-purpose activities feature both science content and pedagogical content knowledge and their relationship to the *nature of science* (NOS) (Abd-El-Khalik, Bell, and Lederman 1998; Aicken 1991; Allchin 2004; Bell 2008; Clough 2004; Cromer 1993; Lederman 1992, 1999; Lederman and Neiss 1997; McComas 1996, 1998, 2004; NSTA 2000; Wolpert 1992) and the *nature of teaching and learning* (APA 1997; Armstrong 2000; Banilower et al. 2008; Bransford, Brown, and Cocking et al. 1999a, 1999b; Brooks and Brooks 1999; Bybee 2002; Cocking, Mestre, and Brown 2000; Donovan and Bransford 2005; Gardner 1999; Goleman 1995; McCombs and Whisler 1997; Michael and Modell 2003; Michaels, Shouse, and Schweingruber 2008; Mintzes, Wandersee, and Novak 1998, 2000; Tobias and Duffy 2009).

Introduction

Organizational Structure of the Book

This book's approximately 80 interactive experiential learning activities are clustered into three sections (i.e., a total of five activities in sections 1 and 2; section 3 has eight 5E mini units that each contain three or more distinct activities, and about 42 Extension activities). Additionally, four appendixes provide background information on the underlying pedagogy.

Section 1. Welcome Back to Interactive Teaching and Experiential Participatory Learning

As in the first two books in this series, the first section of this book is designed to introduce the book's principle pedagogical questions, instructional approaches, and underlying assumptions about teaching and learning. The assumptions about teaching and learning include the following ideas:

1. Intelligence is not primarily genetically determined and limited, and therefore the learning environments within classrooms and the cultures of schools matter (Nisbett 2009).

2. Optimizing learning in schools depends on intelligent, research-informed curriculum-instruction-assessment (CIA) systems that are intentionally designed and cognitively and emotionally motivational, and that feature sense-making interactions between students, teachers, engaging phenomena, and the big ideas of the discipline.

3. Discrepant-event science activities and analogies are powerful instructional tools that align with these assumptions (O'Brien 1991, 1993, 2010).

4. Teachers' professional development related to both science content and pedagogical content knowledge should reflect and model what we know about learning (NCTAF 1996, 1997; NRC 1996, 2010b; O'Brien 2010; PCAST 2010).

The first activity ("Science and Art") is discrepant in that it challenges learners (with words and optional but highly recommended visuals and simple discrepant-event demonstrations) to question how science is similar to—rather than completely different from—art. Accordingly, the activity reintroduces the idea of "science as a way of knowing" that was explicitly featured in section 2 of *More Brain-Powered Science*. The first Extension activity challenges teachers and their students to brainstorm a rationale for the national goal of science for all Americans (Rutherford and Ahlgren 1991) and consider the corresponding implications for the kind of science CIA most appropriate to achieve the desired objectives. As such, the Extension serves as an anticipatory set for the book's longest section (section 3), which focuses on 5E Teaching Cycle–based mini units. Also, as one of only two mixer activities in this book, the activity is designed to help develop a collaborative culture between science teaching and learning as a "team sport." The second activity ("Acronyms and Acrostics Articulate Attributes of Science [and Science Teaching]") can be used with an optional chemistry-related discrepant-event demonstration to make students and teachers more aware of their prior conceptions and implicit beliefs about the nature of science and teaching. "Acronyms and Acrostics" also introduces a minds-on mnemonic strategy as a way to help students learn how to remember scientific terms. As such, Activity #2 can be used as a playful introduction to section 2's focus on reading as an inquiry-oriented process. Additionally, the first Extension activity invites teachers to question the answers of their prior teaching and learning experiences as a framework for yearlong face-to-face or electronically facilitated professional development.

Section 2. Reading, Student Construction of Meaning, and Inquiry-Oriented Science Instruction

For many teachers, the inclusion of a section (activities #3–#5) on reading in a book series that features phenomena-centered, minds-on inquiry-based activities may seem discrepant in and of itself. The following passage from *A Framework for Science Education: Preliminary Public Draft* (NRC 2010a) provides a compelling rationale for this section:

> Researchers have demonstrated the centrality of reading to the practice of science, showing that on average, scientists read for 553 hours per year or 23% of total work time. When the activities of speaking and writing are included as well, the scientists in their study spent on average 58% of their total working time in communication or working in the coordination space … Thus the dominant practice in science and engineering is not "hands-on" manipulation of the material world but rather a "minds-on" social and cognitive engagement with ideas, evidence and argument. Reading, for instance, is an act of inquiry into meaning—an attempt to construct sense from the multiple forms of representation used in science—words, symbols, mathematics, charts, graphs and visualizations [ch. 5, p. 6] …

Introduction

> Being literate in science and engineering requires the ability to construct meaning from informational texts … Reading and interpreting those texts is a fundamental practice of science. Any education in science or engineering needs, therefore, to develop students' ability to read and to produce written text [ch. 5, p. 20] … As such every science or engineering lesson is a language lesson. [ch. 5, p. 21]

The validity of this passage notwithstanding, several common, pedagogically problematic misconceptions suggest this section on reading and science is warranted:

1. By fifth grade, most students have (or should have) successfully transitioned from learning to read to reading to learn.

2. Reading science textbooks is similar to reading other textbooks, fictional stories, or popular print-based media and does not require any special word-decoding skills and information-processing or metacognitive strategies.

3. Teaching students how to read is the sole responsibility and expertise of English language arts and reading teachers.

4. Reading about science is the antithesis of doing "real" inquiry-based science.

5. Reading to learn and learning how to read science textbooks are not essential, FUNdaMENTAL* components of grades 5–12 science classrooms.

Section 2 is designed to challenge these misconceptions with fun minds-on, inquiry-oriented activities that teachers can experience as learners and use with their own grades 5–12 students. Learning how to negotiate one's way through the "language labyrinth" can and should be a playful part of learning science.

Section 2 helps set the stage for section 3 in that brief inquiry-oriented reading passages (e.g., history-of-science case studies and present-day science-in-the-news articles) drawn from internet sites, popular science-type journals, and other science news sources can be used as a complement to phenomena-based discrepant-event Engage- and Explore-phase activities. These kinds of passages raise interesting questions (rather than offer answers) and help establish a motivational need-to-know context and problem-solving situation that demands further investigation. More conventional science textbook passages can be productively used both as in-class, guided reading activities and out-of-

---

* The interactions between teachers, learners, and FUNomena should be FUNdaMENTAL in two senses of the word. First, they should be both emotionally engaging (fun) and cognitively stimulating (mental). Second, they should develop core scientific concepts ("big ideas") that serve as the theoretical and conceptual foundations (or "forest") for contextualizing additional related concepts (or "trees")

NATIONAL SCIENCE TEACHERS ASSOCIATION

class homework assignments during the Explain and Elaboration phases. And of course, reading is a necessary prerequisite for most kinds of performance during the Evaluation phase. As learners become more skilled as readers, their motivation and ability to pursue independent reading outside of school helps them become more scientifically literate lifelong learners. Science-content-based literacy skills are essential components of an intelligent curriculum-instruction-assessment system. Although not a focus of this book, developing students' literacy skills as writers who can clearly and creatively communicate their evolving understanding of science is equally important (e.g., Harris Freedman 1999; Klentschy 2010; Norton-Meier et al. 2008).

Section 3. Integrated Instructional Mini Units: 5E Teaching Cycles

This final section is designed to reintroduce and provide models of 5E Teaching Cycles that integrate CIA. Appendix B in *Brain-Powered Science* (O'Brien 2010) gave an overview of this powerful CIA model that was developed by Biological Sciences Curriculum Study (BSCS) in the late 1980s as an expanded version of the 1970s Science Curriculum Improvement Study's (SCIS) learning cycle. Readers who have not read *Brain-Powered Science* are encouraged to download a synopsis from the model's original designers (Bybee et al. 2006) or simply study the sample activities in this section. The 5E Teaching Cycle of Engage, Explore, Explain, Elaborate, and Evaluate can be used at multiple levels of CIA design, from an individual lesson to a mini unit of a couple of days to a longer unit of between one and four weeks. Though individual lessons can be framed loosely in terms of this sequence, the model takes on its greatest power when used at the longer unit level, which includes an intentional, varied sequence of different kinds of instructional activities spread out over time. Given the purposes and space limitations of this book, the 5E mini units in this final section model the intermediate level of an instructional block of between several days and a week.

Section 3 has a stronger emphasis on big-picture biological, Earth, and space science concepts than any other section in the three-book series. Each 5E mini unit (activities #6–#13) includes several distinct but intentionally sequenced linked activities, and as a result each unit is longer than most other activities in the series. However, it is important to clarify what these 5E mini units are not. Specifically, they are not designed as comprehensive, ready-to-implement curriculum units with detailed daily lesson plans, student reading materials, and handouts. Commercial, grade-level-specific science text-

Introduction

books typically come bundled as seemingly teacher-proof packages. Such packages can be quite useful if designed properly, but even the best of such packages need to be adapted, implemented, and revised "on the front lines in the trenches" of individual classrooms, where professional teachers make informed decisions about how to best serve their own students.

The purpose of these 5E mini units is to serve as a professional development tool to catalyze teachers toward CIA best practices as informed by cognitive science research. To achieve this objective, each 5E mini unit is explicitly linked to one or more of the big ideas in science (i.e., AAAS *Benchmarks* common themes of system, models, constancy and change, and scale; see Appendix C). Cognitive learning theories emphasize learning and teaching as linked processes of minds-on construction of new and improved mental schema scaffolded by the interactions between teacher and student, student and engaging phenomena, and student and student. As such, any individual discrepant-event activity (or 5E mini unit)—although useful as a lesson (or unit) "idea starter" to teach a particular science concept (or interrelated set of concepts)—has a broader purpose as a *visual participatory analogy* to help teachers rethink and refine other activities and units that are already part of their practice. Teachers learning how to teach science and teachers teaching students how to learn science are iterative and integrated, career-long processes. These mini units attempt to present teachers with a way of synergistically sequencing a series of engaging lessons where the whole is greater than the sum of the parts. As this is one of the take-home messages of this book series, it is appropriate to end this book by focusing on how discrepant events can be combined and sequenced into the 5E model of CIA.

Appendixes

Appendix A: ABCs of Minds-on Science Teaching (MOST) Instructional Strategies
Although inquiry-oriented discrepant-event and analogy-based activities are the primary foci of this book series, a variety of other strategies are modeled as well. Appendix A includes an extensive list of instructional approaches that can be incorporated into interactive, research-informed science curriculum-instruction-assessment (CIA). Individual teachers or groups of teachers are invited to see how many A-to-Z activities they can generate by brainstorming, then compare their list to the strategies they regularly use, with an eye to

- expanding their teaching repertoire to better "reach and teach" their diverse student populations by gaining and maintaining their attention, cooperation, and active minds-on participation; and

- supporting their own continued FUNdaMENTAL growth as lifelong learners of the science and improvisational art of science teaching (Tauber and Sargent Mester 2007).

An accompanying visual features the interaction effects of hands-on/hands-off and minds-on/minds-off learning and teaching.

Appendix B: An Integrated, "Intelligent" Curriculum-Instruction-Assessment (CIA) System

Research and policy statements on science education emphasize that a systems view of curriculum-instruction-assessment (CIA) is needed to close the gap between teaching efforts and learning results as measured on student assessments (NRC 2001b, 2001c, 2006). A Venn diagram is provided to help visualize the two-way interactions between these three interrelated components of effective teaching. The 5E Teaching Cycle is one approach to such an integrated CIA design.

Appendix C: Big Ideas in Science: A Comparison Across Science Standards Documents

This three-column table shows how the *Benchmarks for Science Literacy* (AAAS 1993), *National Science Education Standards* (NRC 1996), and *A Framework for Science Education: Preliminary Public Draft* (NRC 2010a) terminology for the big-picture ("forest") ideas of science align. All three documents emphasize the need to teach these ideas explicitly to contextualize the many individual "trees" of science and thereby promote understanding, retention, application, and transfer of scientific concepts over time. This book series provides a rich experiential context for teachers to learn how to incorporate these themes. Section 3 in this volume explicitly models how to teach for these big ideas across multiday instructional sequences.

Appendix D: Science Content Topics

This appendix (in conjunction with the index) can be used to help locate activities by the featured science content.

Activity Format (For New Readers)

As with the previous two *Brain-Powered Science* books, each dual-purpose science discrepant event/science education professional-development activity has the following standard format: Title, Expected Outcome, Science Concepts, Science Education Concepts, Materials, Points to Ponder, Procedure, Debriefing, Extensions, Internet Connections, and Answers to Questions in

Introduction

Procedure and Debriefing. Many readers will be familiar with this format from the previous books, and new readers can probably understand the format from the headings or by working through one activity, so only a few comments are necessary here. Given this book's focus on encouraging teachers to be learners by directly experiencing "cognitive conflict" (Baddock and Bucat 2008; Chinn and Brewer 1993, 1998), the brief Expected Outcome statement and the relatively short explanations of the Science and Science Education Concepts do not need to be read *before* attempting a given activity. Scaffolded inquiry questions embedded in the Procedure and subsequent Debriefing questions are designed to help teacher-users discover the gist of the underlying ideas by doing the activity and reflecting on the results. Though these questions should also prove helpful when using the activities with grades 5–12 students, they are not intended as "teacher-proof scripts" to be followed. Instead, they should model and catalyze questions that the teachers-as-learners and their students will generate as they interact with the discrepant phenomena. Learner-generated questions are critical to learning as they reflect interest and cognitive engagement and provide formative feedback to both the teachers and the learners themselves (Chin and Osborne 2008).

The Answers to Questions in Procedure and Debriefing are intentionally placed at the end of each activity to encourage teachers to approach their own professional development as an inquiry-oriented, discover-the-answers versus read-the-answers activity. Encountering new activities from the perspective of a learner who doesn't know the answers ahead of time gives teachers valuable insights into the perspective of their own students. In particular, "miss-takes" often highlight science misconceptions that, if not directly confronted, have the power to derail the conceptual change process required in learning (Driver, Guesne, and Tiberghein 1985; Driver et al. 1994; Duit 2009; Fensham, Gunstone, and White 1994; Fisher and Lipson 1986; Harvard Smithsonian Center/MOSART; Kind 2004; Meaningful Learning Research Group; O'Brien 2011, Appendix A; Olenick 2008; Operation Physics; Osborne and Freyberg 1985; Posner et al. 1982; Science Hobbyist; Stepans 1994; Taber 2002; Treagust, Duit, and Fraser 1996; White and Gunstone 1992). Activating, challenging, and modifying or replacing seemingly valid misconceptions parallels the similar process that allows science itself to turn "promising, pregnant problems" (i.e., theory and evidence mismatch due to discrepant events) into the "birth" of new ideas and progress (Grant 2006; Youngson 1998).

Several additional format elements serve as catalysts for ongoing individual or group-based professional development that could continue for one or more years after initially experiencing the main activities as learners and testing them as teachers. The Extensions are brief descriptions of related inquiry activities that are useful for independent follow-up work by teachers as a means of formative self-assessment and to support the development of units that link a series of related activities (e.g., 5E Teaching Cycle–based units). The Internet Connections provide resources for teachers (e.g., professional development links, written descriptions, and Quick Time movies of similar or related discrepant-event demonstrations and computer simulations) that, like the Extensions, are starting points for further explorations. Given the continual flux of information on the internet, some links will change URLs or be dropped over time. However, most of the sites are hosted by universities, professional organizations, museums, online encyclopedias, science supply companies, and other such organizations or companies that tend to have a stable, long-term presence on the web. In addition to their inclusion in the text, these sites can be accessed electronically via an NSTA Press online, hyperlinked resource that will also allow for easy updating. E-learning experiences and resources are an ever-growing venue for teacher professional development and "just in time" instructional resources for teaching science across the K–16 range (NSTA 2008).

Most individual activities can be quickly modeled in 15–20 minutes when used with teachers as *science education visual participatory analogies* or as *model science inquiry lessons*. With instructional time so limited in most professional development settings, the activities are designed to be easy to set up, execute, and clean up. When used as science-inquiry activities with grades 5–12 students, the activities could take up to a full class period (or several days of instruction for section 3's 5E mini units) and would optimally be placed in an integrated instructional unit of related concepts and activities.

Closing Comment

This book series is based on the dynamic interplay between the philosophy of science, psychology of learning, and pedagogy of teaching as informed by the students, teachers, and researchers that I have had the pleasure and privilege of learning with and from during my nearly 50 years as a student and teacher. Being a teacher affords one the opportunity and responsibility to be a lifelong learner in a professional learning community that both "stands on the shoulders of giants" (i.e., previous scientists, philosophers,

Introduction

and educators) and lifts up the next generation to see farther and travel farther. With deep gratitude to the past, the *Brain-Powered Science* series reframes a collection of classic and newer science discrepant events as dual-purpose science inquiry lesson examples—analogies for science teacher education. The objective of this somewhat unique pairing is that teacher-readers can learn from minds-on reflection in and on the same FUNdaMENTAL science experiences they can subsequently use to Engage, Explore, Explain, Elaborate, and Evaluate their own grades 5–12 students. As such, the activities should be dissected and reconstructed by teachers in the cauldrons of classrooms. (*Note:* The generous amount of white space in the books is there for your comments, corrections, connections, and next-step extensions.)

My intention is to provide "food for thought" for individual classroom teachers and teacher educators. However, good food tastes even better when shared. Just as progress in science depends on collaboration across individual laboratories within scientific communities, educational progress requires that teachers share both their successes and "miss-takes" (i.e., lessons learned and/or problems identified) with each other within and across science departments, school districts, and electronically linked professional science education associations. Schools will achieve their potential as learning organizations when teachers are learning as much as students, and when teachers leave such schools, they leave behind stronger, smarter organizations that are prepared to identify and resolve new, interesting, and important problems. Our country and world demand and deserve nothing less than truly progressive education. As John Dewey (1938) so eloquently put it in *Experience and Education*:

> The belief that all genuine education comes about through experience does not mean that all experiences are genuinely or equally educative. Experience and education cannot be directly equated to each other. For some experiences are mis-educative (p. 25) ... Wholly independent of desire or intent, every experience lives on in further experiences. Hence the central problem of an education based upon experience is to select the kind of present experiences that live fruitfully and creatively in subsequent experiences (pp. 27–28) ... Unless a given experience leads out into a field previously unfamiliar no problems arise, while problems are the stimulus to thinking ... growth depends upon the presence of difficulty to be overcome by the exercise of intelligence ... it is part of the educator's responsibility to see ... The new facts and new ideas thus obtained become the ground for further experiences in which new problems are presented. The process is a continuous spiral. (p. 79)

Science Education Topics

As with the two previous books, this book has two foci—*discrepant-event science inquiry activities* and linked *visual participatory analogies for science teacher education*. The following topics are organized to show the sequence of sections by the larger frame science education themes. Appendix D lists the science concepts in alphabetical order. In either case, the book does not need to be used in any kind of strict linear sequence, but rather can be explored on a "need to know and use" basis.

Acronyms Used in Science Education Topics

BBS: Black Box System: A hidden mechanism is explored via observation and testable inferences.

BIO: Biological analogies and applications are specifically highlighted.

HOE: Hands-On Exploration: Learners working alone or in groups directly manipulate materials.

HOS: History of Science: A story, case study, or resource from the history of science is featured.

MIX: Mixer: Learners assemble themselves into small groups based on a specific task.

NOS: Nature Of Science: These activities focus on empirical evidence, logical argument, and skeptical review.

PAD: Participant-Assisted Demonstration: One or more learners physically assist the teacher.

POE: Predict-Observe-Explain: These activities use this inquiry-based instructional sequence.

PPP: Paper-and-Pencil Puzzle: These activities use a puzzle, which is typically focused on the NOS; often a BBS.

STS: Science-Technology-Society: The focus is on practical, real-world applications and societal issues.

TD: Teacher Demonstration: The teacher manipulates a system and asks and invites inquiry questions.

TOYS: Terrific Observations and Yearnings for Science: The activity uses a toy to teach science.

Science Education Topics

Section 1. Welcome Back to Interactive Teaching and Experiential Participatory Learning

Activity	Activity Type	Science Concepts
1. Science and Art: Dueling Disciplines or Dynamic Duo?	PPP/MIX TD optional p. 3	NOS as compared to art diffraction of white light to form a rainbow of colors Extensions: Draw-a-Scientist-Test + Science for All Americans + HOS
2. Acronyms and Acrostics Articulate Attributes of Science (and Science Teaching)	PPP TD optional p. 21	NOS, teaching, and mnemonics acid-base indicators and pH Extensions: Analyzing Assumptions About Learning-Teaching

Section 2. Reading, Student Construction of Meaning, and Inquiry-Oriented Science Instruction

Activity	Activity Type	Science Concepts and Learning Principle Modeled
3. Tackling the Terrible Tyranny of Terminology: Divide and Conquer	PPP TD options p. 37	reading, cognition, and NOS acid-based indicators and/or pressure-volume lung model
4. Inquiring Into Reading as Meaning-Making: Do Spelling and Punctuation Really Matter?	PPP p. 51	reading, cognition, NOS, and POE Extension: Cloze test and inference-making
5. Ambiguous Text: Meaning-Making in Reading and Science	PPP p. 63	prior knowledge, reading, and cognition: empirical evidence, logical argument, and skeptical review in science and reading Extension: Fossil Footprints

NATIONAL SCIENCE TEACHERS ASSOCIATION

Section 3. Integrated Instructional Mini Units: 5E Teaching Cycles

Activity	Activity Type	Science Concepts and Big Ideas
6. Resurrection Plant: Making Science Come Alive!	TD/HOE/BIO/HOS p. 81	characteristics of and requirements for life, plant adaptations, and evolution. Extension: Perplexing Plants and Triops
7. Glue Mini-Monster: Wanted Dead or Alive?	TD/BIO/HOS p. 99	characteristics of, requirements for, and scale of life (microscopy)
8. Water "Stick-to-It-Ness": A Penny for Your Thoughts	HOE p. 115	POE, cohesion, adhesion, and surface tension of water Extensions: DHMO pseudoscience Oobleck and non-Newtonian fluids
9. Burdock and Velcro: Mother Nature Knows Best	HOE/BIO/HOS p. 135	form/function fitness, engineering design/Velcro, and microscopy
10. Osmosis and "Naked" Eggs: The Environment Matters	TD/HOE/BIO p. 151	osmosis; the cell membrane as the structure that both separates and connects the cell to its environment and measurement skills
11. 5 E(z) Yet pHenomenal Steps to Demystifying Magic Color-Changing Markers	HOE p. 173	physical and chemical changes, chromatography and molecular movement, and acid-base chemistry and pH indicators
12. 5 E(z) Steps Back Into "Deep" Time: Visualizing the Geobiological Timescale	HOE/BIO/MIX/HOS p. 193	NOS, mathematics of powers of 10 scale, models, geological time, evolution, and STS
13. 5 E(z) Steps to Earth-Moon Scaling: Measurements and Magnitudes Matter	HOE/PAD/HOS p. 231	estimation, mathematics, scale, measurement, astronomical distance (within solar system), models, analytical and aesthetic perspectives

Safety Practices

The discrepant-event, inquiry-based experiments in this book include teacher demonstrations, participant-assisted demonstrations, paper-and-pencil puzzles, and student hands-on explorations. In all cases, it is essential that teachers model and monitor proper safety procedures and equipment and teach students pertinent safety practices through both words and actions. Though the hands-on experiments typically only use everyday materials and household, consumer-type chemicals (e.g., water, sugar, salt, ammonia, and rubbing alcohol), teachers should consider their students' ages and particular teaching environments when deciding how to use particular activities and which safety precautions are necessary. Professional prudence, preparation, and practice greatly reduce the probability of accidents. Effective classroom management and safety are non-negotiable components of effective science teaching even when using very low-risk activities such as those featured in this book. Beyond these activities, consider incorporating the following best-practice safety precautions into your science teaching.

1. Always review Material Safety Data Sheets (MSDS) with students relative to safety precautions in working with hazardous materials. Chemicals purchased from science supply companies come with MSDS. These are also available from various online sites (e.g., *www.flinnsci.com/ search_msds.asp*).

2. Wear protective gloves and aprons (vinyl) when working with hazardous chemicals.

3. Wear indirectly vented chemical-splash goggles when working with potentially hazardous liquids or gases. When working with solids such as soil, metersticks, glassware, and so on, safety glasses or goggles can be worn.

4. Do not eat or drink anything when working in a laboratory setting.

Safety Practices

5. Consider student allergies and medical conditions (e.g., latex, peanut butter, and asthma) when using activities that could elicit a negative response.

6. Wash hands with soap and water after doing the activities dealing with hazardous chemicals or other materials.

7. When working with volatile liquids, heating or burning materials, or creating flammable vapors, make sure the ventilation system can accommodate the hazard. Otherwise, use a fume hood.

8. Immediately wipe up any liquid spills on the floor—they are slip-and-fall hazards.

9. Teach students that the term *chemical* is not synonymous with *toxic*, that *natural* is not synonymous with *healthy and safe*, and that chemicals they encounter on a daily basis outside of the science lab should be used in an informed manner. Scientifically literate citizens and consumers steer between the extremes of chemophobia and careless use of chemicals.

10. Science teachers should stay current on safety threats, environmental risks, and appropriate precautions as part of their ongoing, career-long professional development (Flinn Scientific; Kwan and Texley 2003; Texley, Kwan, and Summers 2004).

References

Council of State Science Supervisors (CSSS). Science Safety Guides (free downloads): *www.csss-science.org/safety.shtml*.

Flinn Scientific, Inc. Safety resources: *www.flinnsci.com/Sections/Safety/safety.asp*.

Kwan, T., and J. Texley. 2003. *Exploring safely: A guide for middle school teachers.* Arlington, VA: NSTA Press.

Texley, J., T. Kwan, and J. Summers. 2004. *Exploring safely: A guide for high school teachers.* Arlington, VA: NSTA Press.

U.S. Department of Health and Human Services: Household Products Database: *http://householdproducts.nlm.nih.gov*.

Section 1:
Welcome Back to Interactive Teaching and Experiential Participatory Learning

Science and Art
Dueling Disciplines or Dynamic Duo?

Expected Outcome

A mixer activity (supplemented by "scientific" art, music, and optional demonstrations) is used to catalyze a conversation on the similarities and differences between the sciences and the arts.

Science Concepts

Learners explore the nature of science (NOS) as an inquiry-based discipline that subjects its creative works to the ongoing tests of empirical evidence, logical argument, and skeptical review (NRC 1996; NSTA 2000). The NOS combines individual competition within and collective collaboration across a community of diverse scientists whose "race to raise and resolve questions" transcends the boundaries of national geography, race and ethnicity, and particular periods of human history. Science can be considered the ultimate of the humanities, where even the greatest work of art is refined and improved upon by later artists (e.g., consider Newton's analogy of "standing on the shoulders of giants" and scientific lineages such as Archimedes → Galileo → Newton → Einstein). Optional, recommended discrepant demonstrations can be used to introduce the science and art of the diffraction of white light into a rainbow of colors, optical illusions, and/or acid-base indicators.

Science Education Concepts

The collaborative, communal nature of science is modeled in the mixing aspect of the activity as the learners collectively analyze and synthesize the information on their individual cards. This visual participatory analogy is also *discrepant* as learners discover that science shares many characteristics and, at times, a common focus with the arts (e.g., Newton's discoveries with prisms related to the nature of color and light, modern-day research on optical illusions, and movie computer graphics). One critical difference between the two disciplines is that the products of science (e.g., naturalistic models and theories that have explanatory and predictive power) are never finalized; they are ongoing works in progress that are always "under construction." The Extension activities invite teachers to brainstorm the rationale for *Science for All Americans*, consider the role of science in history, use the Draw-a-Scientist Test (DAST) to elicit students' often-stereotyped perceptions of scientists, and explore graphic organizers and interactive computer art as visual representations of conceptual understanding. The dual analytical-aesthetic perspective of science is an important yearlong theme that can be supported by emphasizing the big ideas of science, such as systems, models, constancy and change, and scale (see Appendix C).

Materials

Each learner will need a 3 in. × 5 in. index card that contains one of the phrases in the table below. Be sure that the complementary (opposite) card is also in the set of cards that is distributed. If there is an odd number of learners, the instructor can participate. For middle school students, simpler words should be used and/or the more challenging words can be defined on the back of the cards.

systematic and planned	serendipitous and improvisational
methodical and procedural	imaginative and creative
naturalistic and agnostic	may include supernatural and theistic
laboratory- and field-based	studio- and gallery-based
rational arguments	emotional justifications
gradual evolutionary change	rapid revolutionary change
fun "work"	hard "play"
factual and objective	fanciful and subjective
practical	aesthetical (or beautiful)
studies objects' forms	studies objects' functions
follows a single method	uses a variety of methods
quantitative measurements	qualitative patterns
skeptical review	open-minded review
uses conceptual models	uses physical models
individual, solo work	collaborative, communal work
logical-mathematical	visual-spatial
empirical and data-driven	theory-driven
hands-on activity	minds-on activity
communicates to the masses	communicates to the elite
holistic, big-picture view	atomistic, reductionist view

Optional but Recommended Teacher Demonstrations
- An art-of-science slide show (see Extension #5 and Internet Connections for visuals)

- A clear, colorless plastic cup of water on an overhead projector, a prism, or a diffraction grating can be used to display the rainbow of colors "mixed and hidden" in white light.
- Fractals or optical illusions (see Internet Connections: Fractals, Michael Bach, Visual and Auditory Illusions or *Brain-Powered Science*, Activity #5)
- Music that blends animal and/or inanimate sounds of nature and human instrumentalists
- Video clips of performances of "scientific music" such as ArcAttack! (see Internet Connections or search YouTube for other examples)
- Acid-base indicator "magic" signs (see activities #2 and #11 in this book or Activity #12 in *More Brain-Powered Science*)

Procedure

1. Give each learner a 3 in. × 5 in. index card that contains a word or phrase that describes some attribute of science and/or art, and ask the students to "mix and mingle" so each student finds a peer whose card has a word or phrase that seems to be his or her card's opposite. The serial nature of the brief exchanges should allow everyone to find their opposite by process of elimination. Before, during, or after the learners are moving around the room, use one or more of the optional but recommended teacher demonstrations cited in the Materials section. The objective is to present a combined social, visual, and auditory discrepant event that challenges learners' simplistic, mutually exclusive categorizations of science and art. Depending on which of the optional activities are used, this activity can be part of the Engage phase for units on the nature of science, light and optics, and acid-base chemistry.

2. After all the dyads have formed, each learner should have a brief conversation with his or her opposite to consider the following questions:

   a. Does one of the paired cards better describe science (versus art), or do both cards describe attributes of both disciplines?
   b. Considering the supplemental visual and/or auditory representations, are there elements of art within science and science within art?

Points to Ponder

Study the science of art, the art of science. Develop your senses—especially learn how to see. Realize that everything connects to everything else … The most praiseworthy form of painting is the one that most resembles what it imitates.

—Leonardo da Vinci, Italian artist, scientist, engineer, and polymath (1452–1519)

The sciences and arts are not cast in a mold, but formed and shaped little by little, by repeated handling and polishing.

—Michel de Montaigne, French essayist (1533–1592)

There was far more imagination in the head of Archimedes than in that of Homer.

—Francois-Marie Arouet de Voltaire, French writer and philosopher (1694–1778)

Science and art belong to the whole world, and the barriers of nationality vanish before them.

—Johann Wolfgang von Goethe, German poet and novelist (1749–1832)

The most beautiful thing we can experience is the mysterious. It is the source of all true art and all science. He to whom this emotion is a stranger, who can no longer pause to wonder and stand rapt in awe, is as good as dead: his eyes are closed … If I were not a physicist, I would probably be a musician. I often think in music. I live my daydreams in music. I see my life in terms of music. ... I get most joy in life out of music. ... All religions, arts and sciences are branches of the same tree. All these aspirations are directed toward ennobling man's life, lifting it from the sphere of mere physical existence and leading the individual towards freedom.

—Albert Einstein, German American physicist (1879–1955)

c. What are the most unique defining characteristics of science?

3. List (or project) the dyads' pairings of phrases on the board (or screen) under these three headings: Science, Science and Art, and Art. You could also use a Venn diagram to show the unique contrasting versus shared attributes of the two disciplines. Share and briefly discuss one or more of the Points to Ponder quotes to further elaborate on the parallels and differences.

Debriefing

When Working With Teachers and Students

Discuss how the sciences and arts, if superficially considered, appear to be diametrically opposed disciplines with their own set of mutually exclusive attributes. However, if examined more closely, the two disciplines have significant overlap (as visually represented by a Venn diagram). There is no need to try to reach consensus about specific right and wrong answers, but you should focus on cases of pairings that elicit the most discussion. Both artistic and scientific sensibilities and inquiry and representational skills are needed to fully appreciate either discipline as a "consumer" and/or "creator/producer." Overly simplified, neurologically erroneous left-brained (science) versus right-brained (art) analyses misrepresent the interactive, unified nature of the human brain's capabilities (Levitin 2006).

The *National Science Education Standards* (NRC 1996) focus on science as a field of inquiry based on empirical evidence, logical argument, and skeptical review. As with the arts, the motivations of individual scientists are often deeply personal and subjective. But the test of provisional truth in science is what works best in terms of explaining and predicting the broadest range of natural phenomena in the most parsimonious way. This is a more objective standard than informed but more subjective opinions about what makes great art.

When Working With Teachers

(Open-ended discussion questions; see sample answers below)

1. To what extent can art instruction be more scientific? How can scientific principles, skills, and habits of mind be taught through art activities?

2. To what extent can science instruction become more artistic and humanistic? What art and aesthetic skills, principles, inquiry processes, and habits of mind can be taught through science activities? How can the science behind art be used as a motivational tool to attract greater student interest in science? How can artistic skills help students communicate their understanding of science via multiple types of representations?

3. Do your current science curriculum, instruction, and assessment overemphasize certain aspects of science at the expense of their complementary opposites from the above list?

4. Does understanding the science behind nature's artwork enrich or diminish our aesthetic appreciation? Is a rainbow more or less beautiful if we understand the underlying physics?

5. How do the individual daily lessons within a given science unit combine and create synergy where, like all good science and art, the whole is greater than the sum of the parts? Do the ideas of synergy and emergent properties also apply to the relationship between individual units and the entire course and individual courses across kindergarten through grade 12?

Extensions

1. *Science for All Americans?* Challenge teachers (and/or high school students) to brainstorm a list of reasons to convince skeptical legislators, taxpayers, parents, or students that they should care about and invest in equitable and excellent science education as suggested by this bumper-sticker-type slogan (and the title of the AAAS Project 2061 book by James Rutherford and Andrew Ahlgren). Consider that science courses have relatively higher initial and recurring costs (e.g., setup and maintenance of labs), safety and disposal issues, and potentially controversial topics (e.g., transmission of sexual diseases, evolution, and other science-technology-society [STS] issues) than other school subjects.

   Since the 1800s, when science was initially considered a "newcomer" in the long-established classical model of education, various arguments have been advanced to establish the

importance of science in K–12 curricula (DeBoer 1991). John Dewey anticipated many of today's key arguments in books such as *How We Think* (1910), *Interest and Effort in Education* (1913), *Democracy and Education* (1916; see especially chapter 17, "Science in the Course of Study," in the Internet Connections section), and *Experience and Education* (1938). Since then, documents such as NSTA's *What Research Says to the Science Teacher (Project Synthesis)* (Harms and Yager 1981), *A Nation at Risk* (NCEE 1983), *Benchmarks for Science Literacy* (AAAS 1993), the *National Science Education Standards* (NRC 1996), *Why Science?* (Trefil 2008), and *A Framework for Science Education: Preliminary Public Draft* (NRC 2010a) have argued that science education should be designed to do the following:

a. Promote basic scientific literacy for making decisions and taking actions related to personal and social STS issues in our democratic, sci-tech society.
b. Develop intellectual habits of mind and problem-solving skills and attitudes, including both creative-speculative and critical-skeptical thinking (e.g., how to debunk pseudoscience).
c. Maintain economic security and prosperity at both the individual and national levels in a highly competitive, globally interdependent marketplace (e.g., STEM-related jobs and careers and economic equity).
d. Enhance knowledge of and appreciation for the natural world, including both its beauty and inherent value (i.e., aesthetics, avocational interests, and environmental ethics).
e. Understand the story of the human species and our historical, multicultural heritage.
f. Prepare students for subsequent academic courses.

The federal No Child Left Behind legislation has added a seventh reason:
g. Help students achieve on state-mandated standardized exams.

The answers to the "Why science?" question (or "Why do we need to know … ?") have implications for what kinds of science curriculum-instruction-assessment (CIA) are best suited to achieve the desired objectives in particular educational contexts. Section III of this book focuses on 5E Teaching Cycles as an example of an intelligent CIA model.

2. *Draw-a-Scientist Test (DAST):* Decades of research using the DAST have revealed the dominance and persistence of stereotyped images of scientists as unattractive, asocial, bespectacled, older, Caucasian males working in laboratories on potentially dangerous physical science experiments. The DAST can unearth such images so they can be exposed as being overly restrictive in their representations of scientists. It can also reveal the range of students' artistic skills and inclinations. Internet mapping programs and GPS units require both a starting location and a targeted final destination to design an efficient route to connect the two. Similarly, knowledge of students' beginning-of-year attitudes is a necessary prerequisite to developing their scientific literacy. Information on this integrated art and science formative assessment can be found through the following resources:

- American Physiological Society/APS Education Online. DAST Scoring Rubric. *www.the-aps.org/education/2006rts/pdf/DASTRatingRubric.pdf.*
- Barman, C. R. 1996. How do students really view science and scientists? *Science and Children* 34 (1): 30–33.
- Barman, C. R. 1997. Students' views of scientists and science: Results from a national study. *Science and Children* 35 (1): 18–23.
- Chambers, D. W. 1983. Stereotypic images of the scientist: The Draw-a-Scientist-Test. *Science Education* 67 (2): 255–265.
- Finson, K. D. 2002. Drawing a scientist: What we do and do not know after fifty years of drawings. *School Science and Mathematics* 102 (7): 335–345. *http://findarticles.com/p/articles/mi_qa3667/is_200211/ai_n9160846.*
- Fort, D. C., and H. L. Varney. 1989. How do students see scientists: Mostly male, mostly white, and mostly benevolent. *Science and Children* 26 (8): 8–13.
- Mason, C. L., J. B. Kahle, and A. L. Gardner. 1991. Draw-a-Scientist Test: Future implications. *School Science and Mathematics* 91 (5): 193–198.
- See also Activity #7 in *More Brain-Powered Science* for web links to other student assessment instruments related to the nature of science.

3. *Science as the Most Human and Humane of the Humanities?* The history and philosophy of science and its interactions with art, other disciplines, and historical and cultural developments merit career-long study by teachers. *The Timetables of History: A Horizontal Linkage of People and Events* (Grun and Simson 2005) is a wonderful resource for studying the interwoven tapestry of human knowledge as it evolves in parallel and tag-team-type developments across multiple domains of human inquiry and culture. Especially during the past 100 years, science and its technological applications have improved the quantity and quality of life for billions of people. Sustaining these gains and meeting the related economic, energy, environmental, and human health challenges require a scientifically literate population and a cadre of science specialists.

The human drive to observe, study, and reproduce natural images dates back at least 40,000 years to Paleolithic cave art. The development of highly realistic artistic representations during the Renaissance period (14th–17th centuries) provided an intellectually fertile environment for the birth of new ways of seeing and thinking. Leonardo da Vinci (1452–1519) makes an especially intriguing case study in that he stood at the crossroads of the rise of both modern science and art (see Internet Connections). In contrast to his artwork, his scientific work was largely unknown in his time due to his secretive nature and the lack of a community of science scholars to critique and improve it. The dissemination of printing presses (invented in 1453), geographic voyages of discovery, and the rise of scientific societies over the next 200 years caused a rapid expansion in science and the public perception of its importance. By the time of the birth of the United States, a scientific, experimental mindset was in the air. See, for example, Steven Johnson's book *The Invention of Air: A Story of Science, Faith, Revolution, and the Birth of America* (2008) for an intriguing account of this era as seen through the life of polymath Joseph Priestley. Part of Dr. Benjamin Franklin's (a contemporary of Priestley's) political successes can be linked to his international prestige as the man who both tamed the heavens (with his lightning rod) and even invented a new musical instrument, the armonica (see

Internet Connections and *More Brain-Powered Science*, Activity #16). Similarly, other Revolutionary-era leaders were strongly influenced by their Newtonian, law-bound universe worldviews and beliefs in parallel scientific and social advancement.

C.P. Snow's famous essay on the two cultures depicts how far apart the sciences and humanities had drifted by 1951. Although this cultural divide still exists in the general population, the work of modern-era visual artists (e.g., M.C. Escher and Austine), scientists studying topics such as visual illusions and fractals, and film computer-graphics specialists transcends the artificial separation of science, mathematics, and art (see Internet Connections).

4. *Graphic Organizers: Visual Representations of Science Concepts*: The saying "A picture is worth a thousand words" points to the pedagogical power of visual representations to capture and explain complex ideas such as nested, interrelated webs of scientific concepts and theories. A variety of graphic organizers can be used as forms of curriculum-embedded diagnostic, formative, and summative assessment (see Internet Connections: Graphic Organizers for downloadable templates and software). Concept mapping is the most researched graphic organizer in the context of science teaching (Good, Novak, and Wandersee 1990; Mintzes, Wandersee, and Novak 1998, 2000).

5. *Visualizing Environmental Science Statistics With Art*: The challenge of visually depicting the magnitudes of various kinds of human effects on the environment is creatively met by Chris Jordan's interactive art java applets (Internet Connections) that use a zoom function. Compelling examples of this discrepant art form include lightbulbs, cans, plastic bottles, paper bags, cell phones, toothpicks, packing peanuts, and oil barrels. The dynamic interplay between human sensory perceptions and our mental conceptions is also explored in *Brain-Powered Science*, activities #4–#7.

Internet Connections

• ActionBioscience.org:

> Error and the nature of science (article by Douglas Allchin):
> *www.actionbioscience.org/education/allchin2.html*

Why Should You Be Scientifically Literate (article with links by Robert M. Haizen): *www.actionbioscience.org/newfrontiers/hazen.html*

- American Association for Physics Teachers: Physics Photo Contest (past winner galleries): *www.aapt.org/Programs/contests/photocontest.cfm*

- ArcAttack! Musical/Singing Tesla Coil (music videos): *www.arcattack.com*

- Art Institute of Chicago: Science, Art, and Technology (video lectures, extensive web links, etc.): *www.artic.edu/aic/students/sciarttech/2a.html*

- Austine Studios Polarized Light Art or Polages (includes video clips): *www.austine.com*

- Chris Jordan, Photographic Arts: Running the Numbers: An American Self-Portrait: *www.chrisjordan.com/gallery/rtn/#light-bulbs*

- Clip Art Links for (Science) Educators: *http://sciencespot.net/Pages/refdeskclips.html*

- *Democracy and Education*: 1916 book by John Dewey: Columbia University's Institute for Learning Technologies Digital Text Project: *www.ilt.columbia.edu/Publications/dewey.html*

- Dennis Kunkel Microscopy, Inc. Science Stock Photography: *www.denniskunkel.com*

- Disney Educational Productions: Bill Nye the Science Guy: Light and Color ($29.99/26 min. DVD): *http://dep.disney.go.com*

- Exploratorium, the Museum of Science, Art and Human Perception: *www.exploratorium.edu*

- Fractals: A Creative Integration of Science and Art:

    Chaoscope (freeware fractal generator): *www.chaoscope.org*

    Cynthia Lanius's Unit on Fractals: *http://math.rice.edu/~lanius/frac*

    Exploring Fractals (high school math and science): *www.math.umass.edu/~mconnors/fractal/fractal.html*

    Wikipedia: *http://en.wikipedia.org/wiki/Fractals*

Wolfram MathWorld: *http://mathworld.wolfram.com/Fractal.html*

YouTube (search for fractals; some animations have music): *www.youtube.com*

- Franklin Institute: The World of Benjamin Franklin: See especially the Scientist, Inventor, and Musician links and experiments (listen to and play the armonica): *www.fi.edu/franklin*

- Graphic Organizers:

  Freeology: (free download PDF files): *http://freeology.com/graphicorgs*

  Houghton Mifflin Harcourt Education Place (free downloadable PDF files): *www.eduplace.com/graphicorganizer*

  Inspiration ($ Mac/Windows): *www.conceptmapping.com/productinfo/inspiration/index.cfm*

  Institute for Human and Machine Cognition (free concept mapping software and research): *http://cmap.ihmc.us*

  Learning Point Associates: *www.ncrel.org/sdrs/areas/issues/students/learning/lr1grorg.htm*

  Periodic Table of Visualization Techniques (creative, interactive display of more than 100 types): *www.visual-literacy.org/periodic_table/periodic_table.html*

  Wikipedia: *http://en.wikipedia.org/wiki/Graphic_organizer*

- Leonardo da Vinci National Museum of Science and Technology (includes interactive labs): *www.museoscienza.org/english/leonardo*

- M.C. Escher Official Website: *www.mcescher.com*

- McGill University Laboratory for Musical Perceptions, Cognition and Expertise: *www.psych.mcgill.ca/labs/levitin*

- Michael Bach's 88 Optical Illusions and Visual Phenomena: *www.michaelbach.de/ot*

- Microscopy-UK: (2-3D images): *www.microscopy-uk.org.uk/index.html*

- Museum of Science (Boston): Leonardo: Scientist, Inventor, and Artist: *www.mos.org/leonardo*

- National Public Radio (NPR) Morning Edition's Where Science Meets Art series: *www.npr.org/templates/story/story.php?storyId=4111499*

- Nova Celestia: Astronomical Illustrations and Space Art: *www.novacelestia.com*

- Olympus BioScapes International Digital Imaging Competition (video and slide galleries): *www.olympusbioscapes.com*

- Princeton University's Art of Science 2005 Online Competition Gallery (also 2006 and 2009): *www.princeton.edu/artofscience/gallery*

- Project Zero, Harvard Graduate School of Education (cross-disciplinary learning, thinking, and creativity): *http://pzweb.harvard.edu*

- Science and Art.com (The Astronomy Store's Photo Gallery): *www.scienceandart.com*

- *Scientific American Digital*: 105 Mind-Bending Illusions: What They Reveal About Your Brain: *www.scientificamerican.com/special/toc.cfm?issueid=55&sc=singletopic*

- Sea and Sky Astronomical Art Links: *www.seasky.org/links/skylink08.html*

- The Two Cultures and the Scientific Revolution (1951 lecture and book by C.P. Snow): *http://sciencepolicy.colorado.edu/students/envs_5110/snow_1959.pdf*

    See also synopsis at *http://en.wikipedia.org/wiki/The_Two_Cultures*

- Think Quest:

    Leonardo da Vinci: A Man of Both Worlds: *http://library.thinkquest.org/3044*

    A New Perspective on Science and Art: *http://library.thinkquest.org/3257*

- Visual and Auditory Illusions (with interactive demonstration applets): *www.cs.ubc.ca/nest/imager/contributions/flinn/Illusions/Illusions.html#intro*

16

- WikiPedia: *http://en.wikipedia.org/wiki*. Search topics: Renaissance, Leonardo da Vinci, Isaac Newton (the nature of color and light), Benjamin Franklin (scientist and musician), Thomas Edison (motion picture camera), M.C. Escher, and lenticular printing (art-science illusion of 3D depth, motion, or morphing images)
- YouTube: Science-Music-Nature's Art type videos (a subset of a longer list from Activity #13):

    Beethoven's "Moonlight Sonata" and Moon (photo) montage: *www.youtube.com/watch?v=c1tkBNF0ghU* (2:48 min.; piano)

    Enya's "Shepherd Moons" (3:41 min. title track): *www.youtube.com/watch?v=1iRpa-hi1N4* (with images of the Moon)

Answers to Questions in Procedure, steps #2 and #3

Art and science overlap at the level of the science of how the materials and processes used by artists depend on the chemistry and physics of light, colored pigments, and the interactions of different forms of matter, as well as on the biology of perception and human optics. The majority of the terms could be argued to be attributes of both art and science. Science is unique in its emphasis on judging the validity of its theories in terms of empirical evidence, logical argument, and skeptical review that lead to universally applicable, provisional objective truths. It also is more of a collective enterprise and includes continual revisions of its never-finalized, creative products by an ongoing series of investigations that progressively build on and improve or correct errors in previous work.

Answers to Questions in Debriefing, When Working With Teachers, steps #1–#5

1. Careful observation, accurate recording, and/or manipulation of nature are at the heart of both art and science. Art and science teachers can invite students to explore the science behind

various forms of visual art (e.g., optics, color, and the composition of and changes in the chemical materials; see Activity #11). Also, being able to "see in the mind's eye" objects and processes that are not directly observable (and construct physical, mathematical, or theoretical models) has played a key role in the history of science. Art teachers can directly develop these kinds of visual-spatial thinking skills. It is important to emphasize that both art and science require integrating iterative cycles of creative hands-on activity and critical minds-on thinking (i.e., experimentation) to progressively envision, design, and construct "products." Both disciplines build on and extend prior knowledge and proven techniques and require patience, persistence, and practice to progress from naive novice to able apprentice to master and mentor of others. Science-art connections can be a recurring theme throughout both art and science courses.

2. Science teachers can use various types of 2D visual representations (e.g., computer animations, drawings, graphic organizers, graphs, and posters) and 3D models to intentionally develop students' visual-spatial thinking skills at the same time as they develop their ability to conceptualize and communicate scientific ideas (Mathewson 1999; Mayer 2009a; Newcombe 2010). They can also employ music and drama to create participatory simulations of big ideas such as the kinetic molecular theory (Battino 1979) and to recreate historical debates (Begoray and Stinner 2005). In many senses, science can be considered one of the humanities and has a rich international history that has strongly influenced all the other dimensions of human cultures. Truly, science is for *all* Americans (or better, all human passengers on Spaceship Earth). The NRC's *A Framework for Science Education: Preliminary Public Draft* (2010a) and *National Science Education Standards* (1996) and the AAAS *Benchmarks for Science Literacy* (1993) outline the cross-cutting elements or big ideas that should be intentionally, systematically developed in school science (see Appendix C).

3. Educational research shows that the history, philosophy, and aesthetics of science as "a way of knowing" regrettably tend to be given minimal attention in most science classes. Many activi-

ties in this three-book series are designed to explicitly focus on the nature of science.

4. Most scientists would argue that scientific understanding of nature enriches their appreciation and sense of awe and wonder. Rather than destroying the mysterious, each answer in science leads to new questions that point to previously unknown levels of reality. Carl Sagan's 1980 PBS *Cosmos* series exemplifies this dual analytical-aesthetic perspective on science (as do other classic PBS TV series and books such as Jacob Bronowski's *The Ascent of Man* and James Burke's *Connections* and *The Day the Universe Changed,* as well as Activity #13 in this book).

5. Previous books in this series have discussed and section III in this book features the 5E Teaching Cycle approach to lesson and unit design. Policy documents (NRC 2010a) are calling for articulated K–12 learning progressions that systematically construct more sophisticated models and student understanding of the big ideas in science across multiple grade levels.

Activity 2

Acronyms and Acrostics Articulate Attributes Of Science (and Science Teaching)

What's the word that means "Aid to comprehension & recall of scientific terminology & inquiry-based concepts"?

Expected Outcome

Learners' ideas about the nature of science, school science, and science teaching are elicited by their creation of acronyms or acrostics that define key characteristics of science and teaching. An Extension activity provides discussion questions for teachers to explore as part of their ongoing critical reflection and professional development.

Science Concepts

Science vocabulary, mnemonics, and the nature of science (NOS) and science teaching are explored via wordplay. Scientists often have to coin new words to describe new discoveries or inventions (i.e., "foreign words for foreign things"). Acronyms commonly replace the use of longer scientific phrases they represent. Examples include **A**cquired **I**mmune **D**eficiency **S**yndrome (and **H**uman **I**mmunodeficiency **V**irus), **A**tomic **M**ass **U**nits, **C**ompact **D**isc (and **D**igital **V**ideo **D**isc), **C**ompact **F**luorescent **L**amp, **D**eoxyribo**N**ucleic **A**cid (and **R**ibo**N**ucleic **A**cid), **G**lobal **P**ositioning **S**ystem, **L**ight **A**mplification by **S**timulated **E**mission of **R**adiation, **L**ight-**E**mitting **D**iode, **RA**dio **D**etection **A**nd **R**anging, **S**elf-**C**ontained **U**nderwater **B**reathing **A**pparatus, **SO**und **NA**vigation **A**nd **R**anging, and **U**niversal **R**esource **L**ocator. If time permits, it is recommended that this wordplay activity be combined with a quick, discrepant optional teacher demonstration of "chemistry magic" based on color changes in acid-base indicators (i.e., the reactions between household ammonia and either goldenrod paper or phenolphthalein; see the Internet Connections; Activity #11 in this book; and Activity #12, "Magical Signs of Science," in *More Brain-Powered Science*).

Science Education Concepts

This activity serves as a playful, beginning-of-year, diagnostic assessment of learners' attitudes about and perspectives on science and teaching. Teachers often unconsciously acquire unarticulated and therefore unquestioned operational definitions of science and teaching from their years of "apprenticeships of observation" as K–16 students (Lortie 1975). These definitions significantly influence their curriculum-instruction-assessment (CIA) plans. When used with teachers, this activity is a means to activate and re-evaluate the lessons learned from their prior experiences as students and to brainstorm attributes of reform-oriented, research-informed, standards-based school science teaching.

Similarly, grades 5–12 students come to each new year of science classes with prior cognitive conceptions and affective attitudes about the nature of science (NOS) that serve as a foundation that teachers can build on or as misconceptions that need to be "unlearned and reconstructed." For example, students may believe that familiarity with scientific terminology is not only necessary but also sufficient for understanding science. This problem notwithstanding, teachers need to teach students how to use the language of science to learn to read and read to learn science (see section II). Acronyms and acrostics are types of mnemonics or memory-enhancing tools that use the first letters of a word to spell out either a single word, multiple words, sentences, or even poems. This activity can also help students learn to develop their own mnemonics or meaning-making tools.

If the optional teacher demonstration is used, it serves as a *visual participatory analogy* for the big ideas of constancy and change, system-level effects, and emergent properties that arise both when different chemicals react and in written text that constructs meaning by a sequence of progressively nested letters becoming words, which become phrases, which lead to sentences, which make up paragraphs, which form chapters.

Materials

- A blackboard or whiteboard or transparency and projection device to capture and display ideas

Optional Teacher Demonstration

- Household ammonia and either goldenrod paper or phenolphthalein indicator

The goldenrod-colored paper commonly available in schools and office supply stores is not an acid-base indicator. The following companies sell goldenrod paper that turns a bright ("blood") red when it contacts bases such as household ammonia, baking soda, or baking powder. Acid solutions such as vinegar, lemon juice, or, with time, the carbon-dioxide in air will reverse the dramatic color change and allow the paper to be reused.

- Educational Innovations: Color Changing Goldenrod (100 pack; SM925; $9.95): *www.teachersource.com* (888-912-7474)
- Steve Spangler Science: Golden Rod Color-Changing Paper (100 Pack with Spray Bottle; WGRP-250; $9.95): *www.stevespanglerscience.com* (800-223-9080)

Safety Notes

- In part b of the optional teacher demonstration, when household ammonia (or ammonia-based window cleaner) is used in a spray bottle, the "magical sign" should be 2 m from the students and the teacher may wish to wear chemical-splash goggles.

- As of the release of this book, a 1–2% phenolphthalein-alcohol solution is the most common acid-base indicator used in school laboratories. Like all alcohol solutions, it is flammable. Also, after decades of use as the main ingredient in laxative tablets, phenolphthalein was removed from this drugstore product as a suspected carcinogen *when consumed*. If used prudently by the teacher as reusable, color-changing "ink" (i.e., colorless in acids and pink in bases), it seems to involve minimal risk. As with all chemicals, Material Safety Data Sheets (MSDS) supplied by the manufacturers provide an up-to-date source on safety.

Points to Ponder

If I set out to prove something, I am no real scientist—I have to learn to follow where the facts lead me—I have to learn to whip my prejudices.

—Lazzaro Spallanzani, Italian physiologist and microbiologist (1729–1799)

I have steadily endeavored to keep my mind free so as to give up any hypothesis, however much beloved (and I cannot resist forming one on every subject) as soon as the facts are shown to be opposed to it … I love fools' experiments. I am always making them … It is a fool's prerogative to utter truths that no one else will speak.

—Charles Darwin, English naturalist and evolutionist (1809–1882)

In science the primary duty of ideas is to be useful and interesting even more than to be "true." … The mind likes a strange idea as little as the body likes a strange protein and resists it with similar energy. It would not perhaps be too fanciful to say that a new idea is the most quickly acting antigen known to science.

—Wilfred Trotter, English philosopher and scientist (1872–1939)

It is the task of general education to provide the kinds of learning and experience that will enable the student … to apply habits of scientific thought to both personal and civic problems, and to appreciate the implications of scientific discoveries for human welfare … [to] bring the general student understanding of the fundamental nature of the physical world in which he lives and of the skills by which this nature is discerned.

—The President's Commission on Higher Education (1947)

Procedure

1. As a quick introduction to scientific wordplay with acronyms and acrostics, ask the learners to "decode" the component words of one or more of the previously listed science examples and/ or ones taken from popular culture (e.g., **A**utomated **T**eller **M**achine, **C**able **N**ews **N**etwork, **F**requently **A**sked **Q**uestions, **P**ersonal **I**dentification **N**umber, and **U**niversal **P**roduct **C**ode; see Internet Connections for other examples). Mention that in science playing with words in conjunction with systematically playing with new phenomena can be both fun and mentally engaging (see Activity #3 for related activities).

Optional Teacher Demonstrations

This activity can be coupled with either of the following:

   a. *Goldenrod paper demonstration:* Write the letters of an acronym on the colored paper with a black permanent marker. After the learners offer the complete multiword phrase, write the rest of the letters in colorless household ammonia "ink," which will change the goldenrod color to its alkaline red form.

   b. *Phenolphthalein ink variation:* Before class, prepare a large sheet of nonglossy, absorbent white paper by writing the letters of a sample acronym vertically in large block print with a permanent marker, followed by the rest of the letters written horizontally in a colorless phenolphthalein solution that is allowed to dry. Then, in class, after asking the learners to identify the complete phrase, spray the poster with household ammonia. The missing letters will appear in pink, then subsequently fade back to colorless as the ammonia evaporates and carbon dioxide in the air causes the indicator to change back to its colorless form. When dry, this poster can be reused with subsequent classes. See Internet Connections or Activity #12 in *More Brain-Powered Science* for more background information and other variations on magic message demonstrations.

2. Read a standard, "boring" textbook or dictionary definition of science and indicate that you believe students can do better by inventing an acronym (i.e., individual descriptive words) or acrostic (i.e., words linked to form a sentence) for *science*. Ask the

learners to complete the first two steps of a Think-Write-Pair-Share activity where they use the seven letters in the word *science* to either describe its key attributes (when working with students) or describe the kind of science CIA they wish to develop for their students (when working with teachers). If desired, teachers can also be challenged to develop acronyms for the kind of ineffective science CIA that they wish to avoid. (*Note:* CIA is an acronym that reflects the notion that teachers' integrated design of these pedagogical elements is the Central Intelligence Agency of effective teaching.)

3. After the learners individually develop their own ideas, have them first pair up to form dyads and then groups of four to share their ideas and, if possible, develop a single group acronym that captures and integrates the best of their individual ideas to be shared with the entire class.

4. Display and discuss the groups' ideas about the key attributes of science. Note that a diversity of "right" answers is desired to adequately capture the multifaceted nature of science. Also, note if the words teams use reflect an overly narrow or even negative attitude about science.

5. Playfully point out the lengths to which acronyms and acrostics can be taken by displaying the instructor's alliterated answer (see Answers to Embedded Questions). Consider using a little chemistry to "magically" reveal this answer using acid-base indicators as previously described.

Debriefing

When Working With Teachers

If desired and time permits, teachers can be challenged to develop positive and negative acronyms for *teaching* (see examples in Answers to Embedded Questions). Discuss

- how effective science teaching motivates students to "question the answers" they may have mindlessly absorbed from past experiences and personal prejudices. Use the Points to Ponder quotes by Spallanzani, Darwin, and Trotter to emphasize that science is an inquiry-based discipline and that learning science involves continuous conceptual "reconstruction, renovation and remodeling."

- the benefits of teaching students how to develop their own acronyms. Teachers may also wish to further explore the nature of science (see Aicken 1991; Allchin 2004; Cromer 1993; Lederman 1992, 1999; Lederman and Neiss 1997; McComas 1996, 2004; NSTA 2000; Wolpert 1992), quotes from scientists, the use of acronyms and acrostics, and constructivist learning theory (see Internet Connections).

If used, the Optional Teacher Demonstration provides a conversation catalyst to initiate a discussion of the big idea of "the whole is greater than the sum of the parts" with respect to emergent properties in systems, whether as written text or natural phenomena (see Appendix C and Morowitz 2002). If the actual chemistry of the reactions is discussed, it is worth noting that pH is an acronym for power of Hydronium or the negative logarithm (base 10) of the molar concentration of dissolved hydronium ions (H_3O^+). Solutions with pH values below 7 are considered acids and those above 7 are designated as bases (or alkaline). Controlling internal pH values within specific, narrow ranges—despite external environments and foods that can have more varied pH values—is a basic homestatic requirement of all living species (see Internet Connections: PhET and Wikipedia).

When Working With Students

The learners' acronyms or acrostics for *science* provide a playful, informal diagnostic assessment of their prior understanding and attitudes about science, as well as an indication of their familiarity with relevant scientific vocabulary. Avoid leaving the impression that there is a single right answer for this exercise, but rather use the activity to gauge the extent to which students seem both knowledgeable and positive about science. Throughout the school year, encourage students to create their own self-designed mnemonics rather than simply mindlessly memorize those given to them by a teacher or textbook. Student-generated MOMs (minds-on mnemonics) can help them remember when their actual moms aren't around to help (see *Brain-Powered Science*, Activity #12 for other examples). Invite students to share other science mnemonics they've found helpful or challenge them to create a description of the term *acrostic* that is an acrostic

itself (see Answers to Embedded Questions). Also, various types of word puzzles (see Internet Connections: Discovery Education: PuzzleMaker) can be used to engage students in FUNdaMENTAL play with scientific terminology (see also Activity #3).

Extensions (for Teacher Professional Development)

1. *Analyzing Assumptions About Learning and Teaching: Questioning Answers:*

   Teachers can explore various aspects of constructivism and theories of intelligence via the Internet Connections. See Dewey (1938) for a historical context and Tobias and Duffy (2009) for an extended research-informed debate on the relative pros and cons of a range of constructivist to direct instructional strategies. Also, over an extended period of professional development, teachers can individually or collectively examine what their tacit, operational definitions of teaching suggest about the nature of the following:

   Knowledge: Does information = knowledge = understanding = wisdom = ethical behavior?

   a. Is awareness of and/or familiarity with organized, socially shared knowledge (e.g., information in print or digital form) the equivalent of understanding? Can an individual rightly claim to know something that they do not understand? What role does the ability to remember, recall, and regurgitate facts have in educating students for future learning and life? Are lower levels of Bloom's taxonomy (i.e., knowledge and comprehension) prerequisite foundational building blocks for higher-order thinking skills (i.e., HOTS: application, analysis, synthesis, and evaluation)?

   b. Is human knowledge (much less understanding or wisdom) something that can be transmitted, transferred, or passed on between people in a manner analogous to digital data transfer between computers across a WiFi network? If not, does this mean that direct instructional strategies such as lecturing are always inappropriate? Is there a time and place in instructional units for "teacher telling" or "the sage on the stage"? Is being a sage more about providing answers to learners or raising relevant and compelling questions with learners?

Learning: What is "learning" as viewed from biochemical, neurological, psychological, behavioral, and sociological perspectives?

a. To what extent is learning both an inside-out and outside-in interactive process in terms of the mind and its external environment? To what extent is learning an individual versus a social process? In what sense can intelligence be considered "distributed"? Are cooperative learning strategies important preparation for future careers?
b. How does what students already know (or "know that isn't so") affect the process and final product of a teaching and learning experience? How can prior knowledge be a help or hindrance to new learning? Does learning need to include "unlearning" misconceptions?
c. Does "good" teaching always result in student learning? Why or why not?

Students: Are the terms *student* and *learner* synonyms? Do students need to interact with a teacher, other students, and/or actual or simulated phenomena for them to learn?

a. Are students ever truly passive recipients of the teacher's knowledge? Does hands-on activity necessarily imply minds-on learning? Does hands-off activity necessarily imply minds-off?
b. Can a teacher make students learn? Why or why not?
c. How can students use various interactive educational technologies as a substitute for or supplement to a teacher?
d. What are the pros and cons of having a class of students physically together in one location?
e. How is unguided learning from the internet analogous to trying to get a drink from an open fire hydrant? How can standards-based, research-informed curriculum-instruction-

assessment help direct students to use the internet as a learning tool in an intelligent way?

Teachers: Is having a "teacher" a necessary prerequisite of learning? What are the pros and cons of the "teacher" being human, living, and physically present with the student?

a. What roles do "good" teachers play to increase the probability that students will learn more effectively and efficiently than they could without his or her scaffolded assistance?

b. What roles do motivation (teachers catalyzing students' desire to learn) and skill-building (teachers helping students learn how to learn) play in the teaching and learning dynamic?

c. What are the differences between teachers who indoctrinate, inform, instruct, and/or inspire their students?

d. To what extent do (or should) teachers plan for their own obsolescence (for a given student)?

e. How does one's definition of teaching relate to implicit theories about the nature of learning and the purposes of education? How can explicit articulation of these theories affect one's operational definition of "intelligent" curriculum-instruction-assessment (CIA)?

Note: All three volumes in the *Brain-Powered Science* series are designed to support ongoing professional development where teachers' reflections in and on their practice causes them to continually recalibrate their teaching as they re-evaluate and refine their answers to questions such as these.

2. *Historical Hindsight Helps:* Teachers commonly have myopia with respect to the historical antecedents of present-day science CIA policies and practices. DeBoer (1991) presents a historical perspective on the profession with an eye to helping teachers rediscover both the philosophical and political foundations of science education and cases where perhaps the "babies were thrown out with the bathwater" (e.g., compare the 1947 President's Commission Higher Education quote to the 1960s NSF-funded curricula). Modern constructivist learning theories (see Internet Connections) and science standards owe much to their predecessors.

Acronyms and Acrostics Articulate Attributes of Science (and Science Teaching)

Internet Connections

- Acronyms

    Acronym Finder (online dictionary): *www.acronymfinder.com*

    The Free Dictionary: Acronyms: *http://acronyms. thefreedictionary.com*

    WikiPedia: *http://en.wikipedia.org/wiki/Main_Page*. Search: acronym, acrostic, and list of mnemonics.

- *American Educator*, journal of the American Federation of Teachers:

    Winter 2008–09: How Words Cast Their Spell, theme issue on spelling and the Ask a Cognitive Scientist: What Will Improve a Student's Memory?: *www.aft.org/pubs-reports/american_ educator/issues/winter08_09/index.htm*

    Ask a Cognitive Scientist: Students Remember …What They Think About: *www.aft.org/pubs-reports/american_educator/ summer2003/cogsci.html*

- Constructivism and Learning Theories: Supplemental readings for teachers:

    Concept to Classroom: *www.thirteen.org/edonline/concept2class/ constructivism/index_sub5.html*

    Encyclopedia of Educational Technology: *http://eet.sdsu.edu/ eetwiki/index.php/Main_Page*

    Exploratorium: *www.exploratorium.edu/IFI/resources/ constructivistlearning.html*

    Human Intelligence: New and Emerging Theories of Intelligence: *www.indiana.edu/~intell/emerging.shtml#intro*

    Learning-Theories.com: *www.learning-theories.com*

    Research Matters—to the Science Teacher: *http://narst.org/ publications/research.cfm*

    > See, for example, constructivism and the learning cycle, conceptual change teaching, metacognitive strategies, and pedagogical content knowledge

Wikipedia: *http://en.wikipedia.org/wiki/Constructivism* (learning theory/teaching methods)

- Discovery Education: PuzzleMaker: *http://puzzlemaker. discoveryeducation.com*

- Goldenrod Paper as an Acid-Base Indicator:

  Becker Demonstrations: *http://chemmovies.unl.edu/chemistry/ beckerdemos/BD022.html*

  Daryl's Science: *www.darylscience.com/Demos/Goldenrod.html*

  Science Hobbyist: *http://amasci.com/amateur/gold.html*

  Steve Spangler Science: *www.stevespanglerscience.com/ experiment/00000040*

- NSTA Position Statement: Nature of Science: *www.nsta.org/ about/positions.aspx?lid=tnavhp*

- PhET Interactive Simulation: pH Scale: *http://phet.colorado.edu/ en/simulation/ph-scale*

- Wikipedia: pH: *http://en.wikipedia.org/wiki/pH*

Answers to Questions in Procedure, step #2

The first of the following two acrostics for *science* uses multiword alliteration and is consistent with the AAAS *Benchmarks for Science Literacy* (1993) and the NRC *National Science Education Standards* (1996). The second acrostic is clearly inconsistent with these standards.

Positive Acrostic	Negative Acrostic
Skeptical searching with senses for systems	Senseless
Constructed from cooperative	Concepts and calculations,
Investigative inquiries that initiate	Inferior
Earnest exploration of errors about	Exams, and a
Naive notions of nature's narratives via	Note-taking
Controlled and channeled curiosity so that	Cacophony calibrated to create
Empirical evidence from experiments evokes epiphanies	Ennui and exasperation

Acronyms and Acrostics Articulate Attributes of Science (and Science Teaching)

Answers to Questions in Debriefing

When Working With Teachers

Sample acrostics for *teaching* that contrast research-informed best practices of minds-on science teaching (MOST) from ineffective teaching are below.

Positive Acrostic	Negative Acrostic
To	To
Engage and	Exasperate and
Activate attention,	Annoy
Cognitive constructions, and	Children with
Habits of mind and heuristics	Humorless hogwash and
In	Indigestible
Novices	Nonsense that is
Growing	Gainless

When Working With Students

A science education-focused acrostic for the word *acrostic*:

Aid to Comprehension and Recall Of Scientific Terminology and Inquiry-based Concepts.

Section 2:
Reading, Student Construction of Meaning, and Inquiry-Oriented Science Instruction

Activity 3

Tackling the Terrible Tyranny of Terminology

Divide and Conquer

Expected Outcome

An unusually long science word; a simple (i.e., no technical terminology) but incomprehensible paragraph; and verbose, scientific-sounding versions of popular witticisms are used to introduce the science reading skill of word translation via dissection.

Science Concepts

Big, hard words in science are invariably made up of small, easy Greek- and Latin-based prefixes, suffixes, and root words that students can systematically learn, continually use, and creatively recombine. Binomial nomenclature in biology is a special case of this general principle. Part of the nature of science (NOS) is its unique, precise, universally shared language that allows for ongoing professional communications. Optional teacher demonstrations feature color changes with acid-base indicators and the relationship between pressure and volume in the functioning of human lungs.

Science Education Concepts

To learn science is to learn the language of science, but the reverse is not necessarily true. Being able to read, write, speak, and comprehend science terminology is a necessary but insufficient requirement for knowing science. Students need to be sold on the importance of becoming fluent in the initially somewhat foreign language of science. Precision in language leads to precision in thought, experimentation, and reasoned discourse and argumentation about empirical evidence and theories, all of which are necessary elements for learning and contributing to science.

Word analyses have shown that a typical high school science textbook has a significantly greater vocabulary load than a typical foreign language course (Yager 1983). Science also requires students to connect unfamiliar, foreign-sounding words with new, "foreign" concepts that in many cases are not directly accessible through our unaided senses (e.g., atoms, cells, and electromagnetic fields). And in some cases, science uses familiar words (e.g., *work*, *energy*, *power*, and *force*) in very specific ways that differ from their general use in everyday conversations. Teaching students how to read science with understanding involves a multiyear focus on word dissection, grammatical structures, and other strategic reading and metacognitive monitoring skills, set in a context of engaging experiences with science phenomena that generate motivational "need to know" questions. The various parts of this activity are designed to provide teachers and their students with fun and initially discrepant-event experiences related to the nature and role of discipline-specific words in a science classroom. The two subsequent activities (#4 and #5) in this section explore the nature of syntax and context in comprehending the bigger picture painted by paragraphs and longer passages.

Several different analogies are helpful in understanding reading as a process of active construction of meaning. For example, skilled strategic readers use their larger scientific vocabulary base and their ability to rapidly and accurately "identify" previously unknown words ("trees") to efficiently construct an understanding of the main message of a paragraph or extended passage of text (the forest). Conversely, poor readers have to expend so much cognitive effort on identifying individual words (the trees) that they cannot hold enough of the content in their working memory to meaningfully connect to their prior knowledge to construct a sense of the overall "ecology of the forest." Using a microscope analogy, good readers have the ability to switch rapidly between high levels of monocular magnification with restricted fields of view (i.e., the meaning of individual scientific words) and low levels of magnification that have broader, more holistic, stereoscopic fields of view (i.e., paragraph- and passage-level perspectives). If the optional teacher demonstrations with acid-base indicators are used in conjunction with exercise #1, they present dramatic visual analogies for the sense of fear and trepidation that many students have about the "tyranny of terminology" in science classrooms. Converting these negative feelings into a sense of the FUNdaMENTALs of the language of science is an important yearlong theme. Similarly, from an analogical perspective, the lung model suggests that if the pressure placed on novice struggling readers is reduced, the volume of their vocabulary and understanding can be greatly increased.

Materials

The excessively long words, run-on sentences and paragraphs, and scientific witticisms found below may be projected or distributed as handouts. Optional teacher demonstrations that are engaging supplements to exercise #1 include household ammonia and either special goldenrod paper purchased from a science supply house, or phenolphthalein needed for acid-base indicator color changes (see Activity #2 for more information). Also, an operational model and demonstration of the lungs and visual images of volcanic eruptions and diseased lungs can be found via the Internet Connections. (*Note:* A simpler one-lung model only needs an empty 1 L soda bottle, one balloon as a lung, and a second balloon as a diaphragm.)

Points to Ponder

[I] … sometimes employ words new and unheard of, not (as alchemists are wont to do) in order to veil things with a pedantic terminology and make them dark and obscure, but in order that hidden things with no name … may be plainly and fully published.

—William Gilbert, English physician and physicist (1544–1603), in *On the Lodestone and Magnetic Bodies* and *The Great Magnet, the Earth* (1600)

It is impossible to dissociate language from science or science from language.

—Antoine Lavoisier, French chemist and "father" of modern chemistry (1743–1794)

"When I use a word," Humpty Dumpty said in rather a scornful tone. "It means just what I choose it to mean— neither more or less." "The question is," said Alice, "whether you can make words mean so many different things." "The question is," said Humpty Dumpty, "which is to be master—that's all."

—Lewis Carroll, English author and recreational mathematician (1832–1898), in *Through the Looking Glass*

The most powerful thing that can be done is to name something.

—Albert Einstein, German American physicist (1879–1955)

Procedure

When Working With Teachers

Share one or more of the Points to Ponder quotes and/or the misconceptions about the role of reading in science (see Introduction, p. xvii) to initiate a brief teacher discussion on this topic. Then, shift to exercise #1 as a quick role-play of a "serious" grades 5–12 teacher starting a new unit with a "foreign facts before FUNomena" approach.

When Working With Students

Briefly activate students' worst fears about science by semi-seriously introducing a new unit with either or both of the following exercises.

Exercise #1

Science Word Dissection and a Perplexing Paragraph: Project or distribute copies of the following 45-letter-long "discrepant" word as an example of science terms students will need to learn in the next unit. *Optional Teacher Demonstration:* If desired, this word can be made to "magically" appear by using the acid-base indicator or color-change reactions that are described in Activity #2. The long word will appear in bloodred letters when using colorless household ammonia ink on a series of sheets of specially purchased goldenrod paper, or seemingly appear from nowhere in pink letters (i.e., previously colorless phenolphthalein ink) when sprayed with household ammonia (and then slowly disappear as it dries). If either or both of these reactions are used, they can serve as visual analogies to initiate a brief discussion about the "tyranny of terminology" and the short "half-life" of mindlessly memorized words. Also, a simple-to-construct one-lung model can demonstrate the inverse relationship between pressure and volume that accounts for inhalation and exhalation.

Safety Notes
• Indirectly vented chemical-splash goggles, gloves, and apron are required.
• Review MSDS for using hazardous chemicals with students.

Pneumonoultramicroscopicsilicovolcanokoniosis

When this big, "impossible" word causes the expected groans and "you've got to be joking" responses, ask: Do you think your life in this science class would be easier without technical terminology and if we used only easy, small, nontechnical words instead? After briefly considering students' answers to this question, project or distribute copies of the following passage (from Mark Twain's *A Tramp Abroad*) that takes this idea to the extreme:

> So much for one European fashion. Every country has its own ways. It may interest the reader to know how they "put horses to" on the continent. The man stands up the horses on each side of the thing that projects from the front end of the wagon, and then throws the tangled mess of gear forward through a ring, and hauls it aft and passes the other thing through the other ring and hauls it aft on the other side of the other horse, opposite to the first one, after crossing them and bringing the loose end back, and then buckles the other thing underneath the horse, and takes another thing and wraps it around the thing I spoke of before, and puts another thing over each horse's head, with broad flappers to it to keep the dust out of his eyes, and puts the iron thing in his mouth for him to grit his teeth on, uphill, and brings the ends of these things aft over his back, after buckling another one around under his neck to hold his head up, and hitching another thing on a thing that goes over his shoulders to keep his head up when he is climbing a hill, and then takes the slack of the thing which I mentioned a while ago, and fetches it aft and makes it fast to the thing that pulls the wagon, and hands the other things up to the driver to steer with. I never have buckled up a horse myself, but I do not think we do it that way. (Twain [1880] 2004, p. 187)

Ask the learners if the relative absence of long technical words makes this passage easy to comprehend. If not, what would help make it more understandable?

Shift the learners' attention back to the long word and ask them to "dissect" it into smaller, recognizable parts to see if they can piece together the word's meaning. Ask them to look specifically for combinations of letters that appear in other words whose meanings they know and to use these word parts as a conceptual bridge to help "translate" this long word into everyday language.

If desired, the teacher can provide visual context clues by demonstrating a simple low-cost, operational model of the lungs (see Internet Connections) and/or use Google or YouTube images of volcanic eruptions and a dissected cutaway view of diseased lungs. Use this activity as a kick-off for a yearlong activity of regular word dissections with the class. Also, encourage learners to create their own "new" science words. In some cases they'll discover that their "inventions" are already in use, and in other cases maybe they're just "words in waiting" until they're needed.

Exercise #2

Scientific Witticisms: Sophisticated-Sounding Science or Simple Sayings Eschew Obfuscation! That is, you should avoid making concepts obscure or confusing. Or, more simply, be clear and understandable. Throughout the year, periodically give students sayings such as the following to decode and translate from scientific to popular terminology (or the reverse):

a. A singular specimen of the scientific class *Avis* contained within the boundaries of the upper prehensile is equivalently valuable as a doubled inventory of that item located in a low-spreading thicket.

b. Though bryographic plants are typically encountered in substrata of earthy or mineral matter in a concrete state, discrete substrata elements occasionally display a roughly spherical configuration that, in the presence of suitable gravitational and other effects, lends itself to combine translatory and rotational motion. One notices in such cases an absence of the otherwise typical accretion of bryophyta.

c. To possess ocular receptors in the posterior portion of a cephalic protuberance

d. To place a primitive agricultural conveyance in a position anterior to the animal *Equus caballus*

e. To slay a brace of avian creatures with just a single petrous conglomeration

f. Emanating from a culinary vessel into the site of pyrogenic activity

g. To subject a slender illumination device to rapid carbonization on its antipodal points

h. Projecting short, loud noises erroneously toward the top of an arboreal plant

i. To have above normal, average kinetic molecular energy beneath a circular, tight-fitting clothing component

j.   Like sending dense, shelly concretions in front of stout-bodied, artiodacytl creatures

Source: Will Shortz, *Readers's Digest Challenge* (September 2005) and various websites (e.g., see Humorous "Scientific" Laws in the Internet Connections)

Debriefing

When Working With Teachers

Discuss how teachers who have lived in the exotic, "foreign" land of science for many years take its "strange" language and culture as second nature, but that students do not have this advantage and therefore need to be taught how to learn to speak and read the language of science. Instruction in vocabulary building and reading comprehension is an essential complement to (but not a substitute for) hands-on and minds-on experientially based science. Reading is an interactive, constructive process where what the reader already knows (i.e., typically a mix of correct prior conceptions and misconceptions) filters and focuses in a top-down, schema-driven fashion what is perceived and processed from the text in a bottom-up, author-driven fashion.

Science teachers need to connect to and extend their students' prior experiences by bringing relevant phenomena into the classroom via live or virtual models, demonstrations, simulations, and experiments (and by taking students out of the classroom on field trips). Consider, for example, how a model of the human lung and visuals of volcanoes help set a context for the 45-letter word. Engagement with and exploration of science-related experiences create the need both to know science terminology and to use and further develop science reading skills (e.g., in the Explain phase of a 5E Teaching Cycle). Learning to read and reading to learn science can be forms of purposeful, curiosity-driven FUNdaMENTAL minds-on play if introduced at the right time with sufficient cognitive scaffolding. Being fluent in the language of science is a necessary but insufficient condition for understanding the nature and content of science. Use the Points to Ponder quotes to discuss the appropriate balance between doing and reading about science to ensure that students experience the hidden, "foreign" things or FUNomena first before the terminology; how students can be taught how to comprehend and benefit from appropriately timed science reading assignments with an inquiry-based approach in which "wow and wonder come before words"; and the importance of descriptive, precise, universally accepted terminology in science.

When Working With Students

Emphasize your concern with assisting students with not only learning specific terms but also developing skills of learning how to learn science through independent, critical reading. If the 45-letter word is introduced in a unit on the anatomy and physiology of the lungs and respiratory system, be sure to spend time explaining how the inverse relationship between pressure and volume explains the operation of the soda bottle–balloon model of the lungs (see Internet Connections: Lung Model and PhET).

Books for Building Vocabulary

• Terban, M. 2003. *Building your vocabulary and making it great.* New York: Scholastic. Resource book for grades 4–8 students with special attention to prefixes, roots, and suffixes.

• Whitworth, B. 1995. *Building your life science vocabulary.* Belmont, CA: Wadsworth/International Thomson Publishing Company. Helps students overcome their fear of seemingly difficult science terms by providing their root, prefix, and suffix pronunciations; meaning; and examples of usage.

Extensions

1. *Scientific Nomenclature and Nursery Rhymes:* Challenge students to "translate" nursery rhymes into or out of "scientific" language.

   Here is one example:

   > Ahoy, jiggle jiggle! Yonder cavorts a Felis domestica with both a catgut and bow, while quite incredibly a Bos Taurus curveted the luminous globe, causing the diminutive Canis familiaris to cachinnate such a diversion while the trencher eloped with a ladle.

   Source: Detective Shadow. 2000. *Tricky MindTrap puzzles: Challenge the way you think and see.* New York: Sterling Press. This book contains 96 pages of close-up color photography (what is this image?) and logic and wordplay puzzles taken from the popular *MindTrap* games. Other nursery rhymes like this one include Humpty Dumpty (p. 40: "an envelope of albumen, jelly, and membranes ...") and Little Miss Muffet (p. 43: Homo sapiens, arachnid, etc.).

2. *Word of the Week, Cartoon Captions, and Bulletin Board Basics:* Use a bulletin board to feature a science word-of-the-week competition as an extra credit challenge where students prepare miniposters that feature a science word set in the context of a discrepant, humorous, STS, or controversial issue or otherwise engaging science-in-the-news article, cartoon (with the original or their own captions), historical quote, or other source. Winning word entries can be selected by a weighted combination of an expert judge (the teacher) and popular votes (the students) based on factors such as the length of the word, levity of its use, quality of the word dissection, and artistic appeal. The idea is to make dissection and discussion of interesting science words and concepts a FUNdaMENTAL part of the classroom culture.

3. *Crossword Puzzles, Concept Maps, and Comic Strips:* In *More Brain-Powered Science,* Activity #6, Extension #6 describes how to use crossword puzzles as an analogy for science (i.e., the idea of data and theory cross-corroboration or triangulation). Crossword puzzles can also be used in combination with concept mapping as a tool to help students learn science terminology not as isolated words in separate boxes but as hierarchical, interconnected networks of meaning (see Internet Connections: IHMC and Wikipedia). Student-designed comic strips (see MakeBeliefsComix below), the work of professional science cartoonists (see Internet Connections in Activity #12), and Quotations and Humorous "Scientific" Laws (see Internet Connections below) can be used to motivate students to learn and use new scientific terminology in a creative, playful way.

Internet Connections

- A Little Etymology—Greek and Latin Roots—Prefixes and Affixes: *http://ancienthistory.about.com/library/weekly/aa052698.htm*

- Biology Corner: Language of Science: *www.biologycorner.com/worksheets/language.html*

- Discovery (Channel) Education: PuzzleMaker: criss-cross, word search, and other templates: *http://puzzlemaker.discoveryeducation.com*

- Humorous "Scientific" Laws (three websites):

    Unwritten Laws of Life (Huge Rawson): *www.stc.edu.hk/2005/subjects/rs/2000/Stories/laws.htm*

    An Abridged Collection of Interdisciplinary Laws: *www.hdssystems.com/Laws.html*

    Laws of Life: *www.weknowcleanjokes.com/lol-a.html*

- Institute for Human and Machine Cognition: Concept Mapping Tools (free software): *http://cmap.ihmc.us* (Dr. Joseph Novak, the originator of CMapping, is affiliated)

- Just Read Now: Reading in the Sciences (Shared goals, strategies, skills, and research synopses): *www.justreadnow.com/content/science/index.htm*

- Lung Model With Two Lungs and Diaphragm:

    www.scribd.com/doc/3394749/lung-model-with-two-lungs-and-diaphragm-lab

    www.knowledgerush.com/kr/encyclopedia/Bell_jar_model_lung

- MakeBeliefsComix! Make your own comic strip generator: *www.makebeliefscomix.com*

- PhET Interactive Simulations: Gas properties: *http://phet.colorado.edu/en/simulation/gas-properties*

- Quotations for Teaching Science

    Brainy Quote: *www.brainyquote.com/quotes/type/type_scientist.html*

    Dictionary of Scientific Quotations: *http://naturalscience.com/dsqhome.html*

    GIGA Quotes and Witticisms (search by topic or author): *www.giga-usa.com/gigaweb1/quotes2/qutopwitticismsx001.htm*

    Science Quotes: *www.lhup.edu/~dsimanek/sciquote.htm*

- Vocabulary University (online activities including root words): *www.vocabulary.com*

- Wikipedia: *http://en.wikipedia.org/wiki*. Search: biological nomenclature, concept mapping, longest word, laws of science (to contrast to the Humorous Scientific Laws above), and technobabble.

Answers to Questions in Procedure

When Working With Students

Exercise #1

The long word dissects into parts that translate to "a disease of the lungs caused by extremely small particles of silicon-containing volcanic ash and dust." During the 1960s and 1970s, reading research focused on the attention and decoding challenges that long, multisyllabic words and lengthy sentences placed on readers. Various reading formula were developed and grade-level norms were established that allowed teachers to determine the reading level of their textbooks by counting these variables and using charts (rather than actually testing live students). Some textbook publishers revised their texts to the desired readability grade levels by reducing the use of scientific terms and shortening sentences by removing linking words. Unfortunately, later research found that these changes actually increased the cognitive load of science passages by obscuring the meaning that was encoded in science words and making it harder to follow the connected logic of the expository text. A better approach is to teach the skills of word dissection and science content area reading strategies. Also, direct experience with engaging FUNomena, simulations, physical samples, models, and multimedia representations provide powerful context-setting foundations for reading. Doing hands-on, minds-on science can act as a lever for lifting up students' literacy skills and desire to learn to read so they can read to learn.

Exercise #2

a. A bird in the hand is worth two in the bush.

b. A rolling stone gathers no moss.

c. To have eyes in the back of one's head

d. Put the cart (or plow) before the horse.

e. To kill two birds with one stone

f. Out of the frying pan and into the fire

g. To burn the candle at both ends

h. Barking up the wrong tree

i. Hot under the collar

j. Casting pearls before swine

Answers to Question in Extensions

1. Hey diddle diddle, the cat and the fiddle, the cow jumped over the moon; the little dog laughed to see such sport, and the dish ran away with the spoon.

Activity 4

Inquiring Into Reading as Meaning-Making
Do Spelling and Punctuation Really Matter?

Expected Outcome

Learners are asked to read a passage full of misspelled words. Many readers are able to discern the meaning despite the numerous intentionally embedded errors. In a second exercise, learners experience how the meaning of a passage can be dramatically altered if punctuation is changed slightly (while leaving all the words in the same order).

Science Concepts

The human brain is genetically programmed to extract and construct meaning from sensory input. Though some other social organisms have developed a limited ability to communicate through a very restricted range of vocalizations, extensive and ever-evolving languages are a defining feature of all human cultures. Scientists believe that since the dawn of *Homo sapiens*, human babies have literally arrived neurologically "prewired" to naturally learn the semantics, syntax, and grammar of any native language in which they are immersed. (*Note:* If the language skills are not used, infants lose much of this neural-linguistic plasticity through pruning of unused connections.) Written language and reading are far more recent cultural inventions and require formal, intentional instruction to learn how to decipher the code of letters, words, sentences, and paragraphs. As with all sensory stimuli, the brain can be taught to do this more efficiently by focusing on increasing the signal-to-noise ratio by actively looking for and remembering patterns that create meaning and by selectively ignoring extraneous information that is not as relevant.

Science Education Concepts

Science teachers need to teach students how to become more fluent, informed, critical readers of their textbooks and other instructional materials. Reading textbooks for understanding requires going beyond memorizing the bold-printed terms. It demands an active interrogation of text in a manner analogous to the nature of science (NOS), wherein reiterative cycles of empirical evidence, logical argument, and skeptical review result in successive approximation toward tentative truth. Reading depends in part on a minds-on, top-down process in which the reader actively draws on prior knowledge to predict-observe-explain and construct personally relevant meaning. Reading also depends on author-generated syntax and context clues that present bottom-up patterns that help readers extract meaning and make sense of the reading (or discover the "forest for the trees") by recreating an approximation of the author's intended message. This activity examines the process of "reading the lines" or sentences as informed by spelling, grammar, and syntax.

Both of the following exercises serve as *visual participatory analogies* that model how reading, like science, is a minds-on, inquiry-driven, pattern-finding interactive and constructivist process of "making meaning" from what may initially be somewhat puzzling or discrepant information. Skilled, strategic readers select and interpret empirical evidence, form inferences, query the text, and draw conclusions. Similarly, scientists read nature's signs and symbols to develop and test various "stories" (or theories) about how nature works. The history of science contains many fascinating case studies that can be used to help students develop a sense of the storylike nature of science (see Internet Connections: ActionBioscience.org, Centre for Science Stories, and the National Center for Case Study Teaching in Science). Although reading from textbooks is commonly included in the Explain phase of a 5E Teaching Cycle, short inquiry-oriented reading passages may be used to initiate investigations as part of the Engage, Explore, and/or Elaboration phases. Scientists read (and write) before, during, and after experimentation.

Points to Ponder

Most of the fundamental ideas of science are essentially simple, and may, as a rule, be expressed in language comprehensible to everyone.

—Albert Einstein, German American physicist (1879–1955)

Even for the physicist, the description in plain language will be the criterion of the degree of understanding that has been reached.

—Werner Karl Heisenberg, German physicist (1901–1976)

The information I most want is in books not yet written by people not yet born.

—Ashleigh Brilliant, American wordplay humorist (Pot-Shots #3242, 1985)

Materials

The sample passages on this page and page 55

Procedure

Exercise #1

Project (or photocopy) the following passage for all to see and "read."

The Paomnnehal Pweor of the Hmuan Mnid

Cna yuo raed tihs? Olny sum plepoe can. Fi yuo cna raed tihs, yuo hvae a sgtrane mnid too. I cdnuolt blveiee taht I cluod aulaclty uesdnatnrd waht I was rdanieg. The phaonmneal pweor of the hmuan mnid, aoccdrnig to a rscheearch at Cmabrigde Uinervtisy, it dseno't mtaetr in waht oerdr the ltteres in a wrod are, the olny iproamtnt tihng is taht the frsit and lsat ltteer be in the rghit pclae. The rset can be a taotl mses and you can sitll raed it whotuit a pboerlm. Tihs is bcuseae the huamn mnid deos not raed ervey lteter by istlef, but the wrod as a wlohe. Azanmig huh? yaeh and I awlyas tghuhot slpeling was ipmorantt! If you can raed this fsacnitaing tex forwrad it.

Source: e-mail/fax circulation; original source unknown. A similar version is attributed to Udder Buffoonery Productions, LCC at *http://planetperplex.com/en/item152.*

Exercise #2

Distribute a copy of form A (p. 55) of the following passage to half of the class and form B to the other half of the class and have each half complete a Think-Write-Pair-Share cycle:

Think: Individually predict what Adorabelle's reaction would be to the "love note."

Write: Keeping the exact words, change the placement of punctuation marks (e.g., beginning and ending of sentences) to create a note that has the exact opposite emotional tone and meaning.

Pair/Share: Get together with someone on the other side of the room (who has the alternative form of the passage) and compare your work. Discuss the importance of punctuation marks in helping frame the meaning of text and the implications of this exercise relative to constructing meaning from science textbooks and how reading is more than word recognition and decoding.

Form A

How I long for a girl who understands what true romance is all about. You are sweet and faithful. Girls who are unlike you kiss the first boy who comes along, Adorabelle. I'd like to praise your beauty forever. I can't stop thinking you are the prettiest girl alive. Thine, Tyrone.

Form B

How I long for a girl who understands what true romance is. All about you are sweet and faithful girls who are unlike you. Kiss the first boy who comes along, Adorabelle. I'd like to praise your beauty forever. I can't. Stop thinking you are the prettiest girl alive. Thine, Tyrone.

Source: Sobol, D. J. 1985. The case of the angry girl. In *Encyclopedia Brown and the Case of the Mysterious Handprints*, pp. 24–30, 78–79. New York: Bantam Skylark Books.

This short excerpt is drawn from a delightful inquiry-based book series (for grades 3–5 readers) that feature cases that a police captain gives to his son for him to solve using his observation skills and analytical and inferential reasoning abilities. Readers are challenged to solve the mysteries before the end of the story, where the answer is revealed. In essence, Encyclopedia Brown (like all scientists) solves discrepant-event puzzles by relying on empirical evidence, logical argument (e.g., testable inferences), and skeptical review.

Debriefing

When Working With Teachers and Students, Exercise #1

Discuss learners' affective and cognitive reactions to the spelling-error-filled passage. Note that our prior knowledge (schemata) filters, colors, and provides an interpretative focus or "picture frame" for the empirical data we select, collect, and reflect on with our two eyes and "mind's eye." The human brain is designed to extract meaning based on patterns we construct, even if it involves ignoring or reinterpreting some of the "facts" of our observations. This characteristic is both a strength and, at times, a weakness of our thought processes;

to some extent, we "see" what we expect to see. Our internal, often tacit mental theories about what makes sense inform and transform in a top-down fashion empirical data that come to us from a bottom-up direction; this interplay between the mind and its environment is the foundation of science. (*Note:* In *Brain-Powered Science*, section 2, activities #4–#7 feature the relationship between human perceptions and conceptions.) Also, discuss the Einstein and Heisenberg quotes on the "simplicity and clarity" of good scientific theories that enable scientists to "see the forest for the trees" (i.e., unifying themes and big ideas) to explain a broad range of phenomena.

When Working With Students, Exercise #1

Most students will find this exercise to be both funny (i.e., humorous and perhaps a little odd in a science class) and fun (i.e., mentally and emotionally engaging). Use students' foreign language translation success to emphasize that both their brains and science are designed to extract "meaning from messes" (or order, unity, and cosmos from the seeming chaos or disorder of a incredibly diverse but unified universe). Science is "fundamentally" both a FUN and MENTAL activity that students can successfully learn and perhaps even contribute to in a future career (note Ashleigh Brilliant's quote on p. 53). Remind students that reading a science textbook is an inquiry process that requires them to actively pursue answers to questions by engaging in a dialogue of discovery with the author and that correct spelling does really matter in science, especially because so many related words are quite similar in spelling but different in meaning (e.g., *aerobic* and *anaerobic*; *alkane*, *alkene*, and *alkyne*; *exothermic* and *endothermic*; *fusion* and *fission*; *gel* and *cell*; *hypertonic* and *hypotonic*; and *mitosis* and *meiosis*).

When Working With Teachers, Exercise #2

If a picture is worth a thousand words, how much more powerful a tool we have in science teaching when we can also let nature do the talking through phenomena that provide a common experiential foundation to provide context clues to help students literally make sense of readings from the assigned textbook (e.g., demonstrations, experiments, field trips, multimedia, and simulations). Discuss how FUNomena before facts, wow and wonder before words, or real before reading are generally good pedagogical strategies to follow. Also, emphasize that understanding science text involves more than word recognition and decoding (skills to be developed throughout the year using word dissection strategies; see Activity #3).

When Working With Students, Exercise #2

Students with reading difficulties may actually be able to decode individual words reasonably well but fail to make correct use of punctuation marks, syntax, and context clues to monitor and adjust the pace and focus of their reading. If they lack these essential, teachable reading skills, students may lose their way during reading and be unable to stay on task and make sense of or construct meaning from the text. These problems can be identified when students are asked to read aloud short passages (i.e., science teachers may want to arrange private, one-on-one short meetings to make this initial screening assessment). Teaching all students to meet the demands of reading expository science text is the job of every science teacher. Collaborating with reading and special education teachers is always a good idea if necessary, but it is essential to provide additional appropriate, out-of-class literacy support for students with clear reading difficulties. However, these students should never be deprived of the opportunity to learn from direct, minds-on experiences with science FUNomena for remedial reading instruction. Instead, science and reading/literacy instruction should be designed to support each other synergistically. The internet is a rich source of reading material with linked multimedia that may offer greater motivational support and cognitive scaffolding than standard textbooks. Effective science teachers use, but are not limited to, the "two by four" constraints of the two covers of the textbook and four walls of the classroom that might otherwise, by themselves, produce "bored" students.

Extensions

1. *Nursery Rhyme Nuances:* Emphasize that reading is more than deciphering individual words or sentences (i.e., it's reading between and beyond the lines using prior knowledge to make both interpolations and extrapolations) by presenting the following two-sentence paragraphs and asking the learners to infer

their very different meanings. The first sentence, "Mary had a little lamb ...," is followed by

a. Its fleece was white as snow. (Conventional poem; Mary is a little girl who owns a lamb.)
b. She spilled mint jelly on her dress. (The English serve mint jelly with cooked lamb.)
c. It was such a difficult delivery the vet needed a drink. (Mary is the name of a pregnant ewe that just gave birth.)

Source: Altwerger, B., C. Edelsky, and B. Flores. 1989. Whole language: What's new? In *Whole language: Beliefs and practices, K–8,* ed. G. Manning and M. Manning, pp. 9–23. West Haven, CT: NEA Professional Library. ERIC# ED309387. *www.eric.ed.gov*

2. *Playing With Punctuation:* Other humorous takes on the importance of punctuation in ascertaining meaning that can be shared with both teachers and students include

   • Truss, L. 2003. *Eats, shoots and leaves: The zero tolerance approach to punctuation.* New York: Penguin Books. Also available as an audio CD.
   • Phonetic punctuation comedy routines of Victor Borge (Look at *www.YouTube.com* for video clips or *www.Amazon.com* for tapes of his performances.)

3. *Be CLOZE-Minded: Prospecting Power and Mining for Meaning:* The cloze test is a reading-comprehension assessment procedure where students are asked to use both their general and content-specific vocabulary knowledge and inferential reasoning skills to fill in the blanks of a passage that has had every fifth (or seventh or ninth) word removed. The test is useful for evaluating the suitability of a given text for a particular student and the reading comprehension skills of students. Also, teacher-selected cloze test passages can be used to help students learn how to become interrogators or "prospectors" of their science textbook by actively "mining it for meaning." Searching for patterns and creating meaning based on the interaction of focused observations and critical self-assessment of prior knowledge is "doing science," whether the observations (or empirical evidence) come from a textbook or hands-on materials. Intelligently designed curriculum-instruction-assessment (CIA) develops students who are literate in both general and science domain–specific ways by creating a positive synergy between these two ways of knowing.

See Internet Connections: The Fifth Dimension and Wikipedia: Cloze test.

4. *Translating Reading Research Into Practice*: In addition to the Internet Connections, a number of teacher-friendly books have been written to assist science teachers who want to teach their students prereading, during reading, and postreading comprehension strategies:

- Barton, M. L., and D. L. Jordan. 2001. *Teaching reading in science: A supplement to teaching reading in the content areas teacher's manual.* Aurora, CO: McRel. 800-933-2723; *www.mcrel.org:80/topics/products/49.*

- Billmeyer, R., and M. L. Barton. 2002. *Teaching reading in the content areas: If not me, then who?* 2nd ed. Alexandria, VA: Association for Supervision and Curriculum Development. *www.mcrel.org/topics/Literacy/products/11.*

- Douglas, R., M. P. Klentschy, K. Worth, and W. Binder, eds. 2006. *Linking science and literacy in the K–8 classroom.* Arlington, VA: NSTA Press.

- Harris Freedman, R. L. 1999. *Science and writing connections.* White Plains, NY: Dale Seymour Publications.

- Mayer, R. E. 2009. *Multimedia learning.* 2nd ed. New York: Cambridge University Press.

- Santa, C. M., and D. E. Alvermann, eds. 1991. *Science learning: Processes and applications.* Newark, DE: International Reading Association.

- Saul, E. W., ed. 2004. *Crossing borders in literacy and science instruction: Perspectives on theory and practice.* Newark, DE: International Reading Association and Arlington, VA: NSTA Press.

- Their, M., with B. Daviss. 2001. *The new science literacy: Using language skills to help students learn science.* Portsmith, NH: Heinemann.

- Vacca, R. T., and J. L. Vacca. 2007. *Content area reading: Literacy and learning across the curriculum.* 9th ed. Boston, MA: Allyn & Bacon.

- Vasquez, J. A., M. W. Comer, and F. Troutman. 2010. *Developing visual literacy in science, K–8.* Arlington, VA: NSTA Press.

- Wellington, J., and J. Osborne. 2001. *Language and literacy in science education.* Philadelphia: Open University Press.

Internet Connections

- AAAS Project 2061 Textbook Evaluations (criteria and actual analyses of commercial books): *www.project2061.org/publications/textbook/default.htm*

- ActionBioscience: *www.actionbioscience.org/education*

    See Classroom Methodology: Using Case Studies to Teach Science (by Clyde Freeman Herreid) and Tapping Into the Pulse of the History of Science With Case Studies (by Douglas Allchin)

- *American Educator*, Winter 2008–09: "How Words Cast Their Spell" theme issue on spelling; includes PDF of the "Ask a Cognitive Scientist" column: "What Will Improve a Student's Memory?": *www.aft.org/pubs-reports/american_educator/issues/winter08_09/index.htm*

- *American Educator*, Spring 2002, "The Story of Science and the Power of Story" issue: *www.aft.org/pubs-reports/american_educator/issues/spring02/index.htm*

- *American Educator*, "Ask a Cognitive Scientist" column: *www.danielwillingham.com*

    The Privileged Status of Story: *www.aft.org/pubs-reports/american_educator/issues/summer04/cogsci.htm*

    The Usefulness of Brief Instruction in Reading Comprehension Strategies (PDF download): *www.aft.org/pubs-reports/american_educator/issues/winter06-07/index.htm*

    Have Technology and Multitasking Rewired How Students Learn?: *www.aft.org/newspubs/periodicals/ae/summer2010/index.cfm*

- Centre for Science Stories (historical case studies): *http://science-stories.org*

- Center on Instruction (see Reading): *www.centeroninstruction.org/index.cfm*

- Colorado State University: Critical Reading and Cognitive Reading Theory: *http://writing.colostate.edu/guides/reading/critread/index.cfm*

- Cuesta College Academic Support: How to Read Effectively in the Sciences: *http://academic.cuesta.edu/acasupp/AS/621.htm*
- International Reading Association: IRA-published books and journals (e.g., *The Reading Teacher* and the *Journal of Adolescent and Adult Literacy*) with searchable database; nonmembers can access abstracts and members can access full-text articles: *www.reading.org*
- Learning Point Associates (formerly NCREL): *Adolescent Literacy and Reading Strategies* (31): *www.learningpt.org/literacy/adolescent/strategies.php*
- Metacognitive Awareness of Reading Strategies Inventory (MARSI): grades 6–12, 30-item questionnaire: *www.literacyintervention.org/documents/MARSI.pdf* (original source: Mokhtari, K., and C. A. Reichard. 2002. *Journal of Educational Psychology* 94 (2): 249–259).
- National Association for Research in Science Teaching (NARST): *Research Matters to the Science Teacher:* A Guide to Assessing, Selecting and Using Science Textbook Visuals Using Textbooks for Meaningful Learning in Science: *www.narst.org/publications/research.cfm*
- National Center for Case Study Teaching in Science: *http://sciencecases.lib.buffalo.edu/cs*
- National Reading Panel: *www.nationalreadingpanel.org*
- Prereading Strategies: *http://departments.weber.edu/teachall/reading/prereading.html*
- Reading Quest.org: Making Sense in Social Studies (Most of the strategies also apply to science): Strategies for Reading Comprehension (28): *www.readingquest.org/strat/home.html*
- Reciprocal Teaching: *www.ncrel.org/sdrs/areas/issues/students/atrisk/at6lk38.htm*
- The Fifth Dimension: Cognitive Evaluation (Site of Richard Mayer): Cloze Test: *www.psych.ucsb.edu/~mayer/fifth_dim_website/HTML/cloze_test/cloze_home.html*
- Wikipedia: *http://en.wikipedia.org/wiki.* Search topics: cloze test, graphic organizer (links to types), reading process (links to SQ3R), and schema.

Activity 5

Ambiguous Text
Meaning-Making in Reading and Science

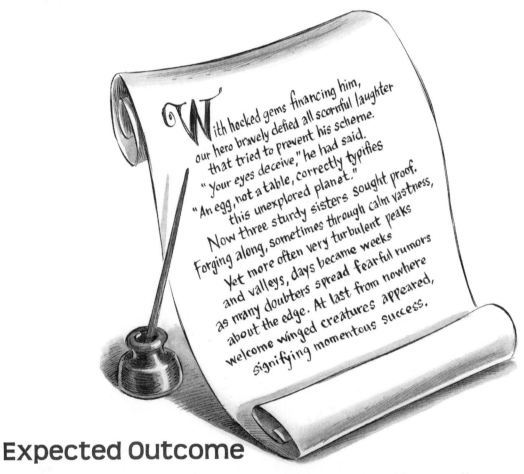

With hocked gems financing him,
our hero bravely defied all scornful laughter
that tried to prevent his scheme.
"Your eyes deceive," he had said.
"An egg, not a table, correctly typifies
this unexplored planet."
Now three sturdy sisters sought proof.
Forging along, sometimes through calm vastness,
yet more often very turbulent peaks
and valleys, days became weeks
as many doubters spread fearful rumors
about the edge. At last from nowhere
welcome winged creatures appeared,
signifying momentous success.

Expected Outcome

Learners are asked to read one or more passages of ambiguous, discrepant text where they understand the individual words (or "trees") but are hard-pressed to connect the words with an overall context (or "forest") to extract and construct meaning from the passage.

63

Science Concepts

Understanding science and comprehending any kind of written text require learners to *collect, select,* and *reflect* on empirical data and use logical argument and skeptical review to construct tentative, provisional truths. Reading a science textbook for comprehension is an inquiry-driven, meaning-making activity in which the reader uses prior knowledge to actively "interrogate" the text by reading not only the individual words and lines but also between and beyond the lines of text. Doing hands-on science and minds-on reading about science both involve making assumptions, raising questions, and drawing inferences whose validity is assessed in light of additional data that is subsequently collected. The gestalt principle that the whole is greater than the sum of the parts applies to science explorations and to learning about science via discussions with authors that include critically reading the story of science. Learning to listen, speak, read, and write science "stories" is part of the work of both scientists and students and an essential part of developing scientific literacy.

Science Education Concepts

Attention, information processing, memory encoding, and transfer of learning are strongly affected by the meaningfulness of the content with respect to learners' prior knowledge and the extent that the new experience activate these pre-existing schemata or conceptual networks. Science as a process of reading facts first and telling, teaching, and testing technical terminology typically fails to activate relevant prior knowledge or motivate student interest and effort. Instead, such miseducation produces a sense of fruitless frustration and meaninglessness about learning science, even for those adept at memorizing and regurgitating decontextualized facts. By contrast, a "FUNomenon first and facts follow" or "wow and wonder before words" approach is more sensible in both meanings of the word (i.e., able to be experienced through the senses and logical). So-called illiterate, primitive people and scientists both learn to "read the world" and so, too, must our students learn to observe, interpret/decipher/ translate, and learn from nature's symbols and signs. Reading difficulties should not be an immovable impediment to students' ability to learn science. To the contrary, phenomena-centered and inquiry-based science should be a means to catalyze and support the development of broader literacy (e.g., listening/speaking, reading/writing, and visual/digital/multimedia literacies) and numeracy skills (e.g., mathematical computation, graphical and statistical analyses, and spreadsheets and data mining).

These ambiguous passages serve as *visual participatory analogies or models* where teacher-learners experience the frustration of reading text that is unintelligible because it does not activate their prior experiential knowledge base or provide sufficient clues to help establish a "big picture" context that makes sense. This experience is analogous to (but in some sense not as difficult as) that which many students experience when they read strange-looking, foreign-sounding words and passages in a science textbook without the necessary experiential foundation to help set the context and create a need to know. Whether a science textbook and other reading materials are perceived by students as a "language labyrinth to languish in" or a "lively laboratory to learn in" depends on when and how teachers require the reading materials to be used relative to the experiential and conceptual scaffolding provided. Students will not learn to read or read to learn science by a passive, osmotic process of immersion in concentrated, text-rich environments. Quality science curriculum-instruction-assessment is not limited by students' prior reading skills, but synergistically develops both science and reading skills as part of scientific literacy.

Materials

- The reading passages can be distributed in paper form or projected.
- *Optional*: Google Images can be used to find a visual cue ("a picture is worth a thousand words") for each of the passages (i.e., as hints *after* students struggle with the text-only experience).

Procedure

1. Distribute individual paper copies or project one or more of the following reading passages (without the corresponding titles or visual context clues). If desired, learners in different sections of the room can be assigned different passages. When working with teachers as the "learners," reserve at least one passage for use in the Extension #2 activity.

Points to Ponder

Read not to contradict and confute; nor to believe and take for granted; nor to find talk and discourse; but to weigh and consider. Some books are to be tasted, others to be swallowed, and some few to be chewed and digested ... Reading maketh a full man ...

—Francis Bacon, English natural philosopher and futurist (1561–1626)

The more that you read, the more things you will know. The more that you learn, the more places you'll go ... Think left and think right and think low and think high. Oh, the thinks you can think of if only you try.

—Dr. Seuss (Theodore Geisel, 1904–1991), from *I Can Read With My Eyes Shut* and *Oh, the Thinks You Can Think!*

Although the Standards emphasize inquiry, this should not be interpreted as recommending a single approach to science teaching ... Conducting hands-on science activities does not guarantee inquiry, nor is reading about science incompatible with inquiry.

—*National Science Education Standards* (1996), p. 23

2. Ask the learners to complete a Read-Think-Write-Pair/Share sequence where they first individually (a) silently read the passage and write down any words that they do not understand, and (b) generate one or more hypotheses about what they think their passage is about (e.g., write down a two- to four-word title for the passage). Then have the learners form dyads or triads to share their hypotheses and (c) discuss why inferring the context or gist of their passage is difficult even when they understand all the individual words, (d) how this experience is similar to or different from reading a science textbook, and (e) what could be done to make the passage more comprehensible.

3. Ask the groups to share their hypotheses (or titles) with the entire class and discuss the logic of the various ideas. If any groups are

having difficulty coming up with potential contexts for a given passage, you may wish to offer additional verbal and/or visual context clues to activate the learners' prior knowledge (e.g., analogies that state "this is sort of like," "this is something someone in your household does at least once a week," or "this represents an important historical event"). When the correct titles and/or the picture answers are shared, they are certain to elicit verbal expressions and body language indicators that "light up" when the learners "get" the puzzle of the passage's meaning. Use the learners' responses to segue to the Debriefing discussion.

Sample Passage #1

A newspaper is better than a magazine. A seashore is a better place than the street. At first it is better to run than to walk. You may have to try it several times. It takes some skill but it is easy to learn. Even young children can enjoy it. Once successful, complications are minimal. Birds seldom get too close. Rain, however, soaks in very fast. Too many people doing the same thing can also cause problems. One needs a lot of room. If there are no complications it can be very peaceful. A rock can serve as an anchor. If things break loose from it, however, you will not get a second chance.

Source: Klein, M. 1981. Activity #36: Context and memory. In *Activities handbook for the teaching of psychology*, vol. 1, ed. L. T. Benjamin and K. D. Lowman, p. 83. Washington, DC: American Psychological Association. For related research, see: Miller, G. A., and J. A. Selfridge. 1950. Verbal context and the recall of meaningful material. *American Journal of Psychology* 63: 176–185.

Sample Passage #2

The procedure is actually quite simple. First, you arrange things into different groups. Of course one pile may be sufficient depending on how much there is to do. If you have to go somewhere else due to lack of facilities that is the next step, otherwise you are pretty well set. It is important not to overdo things. That is, it is better to do too few things at once than too many. In the short run this may not seem important but complications can easily arise. A mistake can be expensive as well. At first the whole procedure will seem complicated. Soon, however, it will become just another facet of life. It is difficult to foresee any end to the necessity of this task in the immediate future, but then one never can tell. After the procedure is completed, one arranges the materials into groups again. Then

they can be put into their appropriate places. Eventually they will be used once more and the whole cycle will then have to be repeated. However, that is a part of life.

Source: Bransford, J. D., and N. S. McCarrell. 1974. A sketch of a cognitive approach to comprehension. In *Cognition and the symbolic processes*, ed. W. Weimer and D. S. Palermo, pp. 189–229. Hillsdale, NJ: Lawrence Erlbaum Associates. Modified by the authors from the original source: Bransford, D., and M. K. Johnson. 1972. Contextual prerequisites for understanding: Some investigations of comprehension and recall. *Journal of Verbal Learning and Verbal Behavior* 11: 717–726.

Sample Passage #3

Every Saturday night, four good friends get together. When Jerry, Mike, and Pat arrived, Karen was sitting in her living room writing some notes. She quickly gathered the cards and stood up to greet her friends at the door. They followed her into the living room but as usual they couldn't agree on exactly what to play. Jerry eventually took a stand and set things up. Finally, they began to play. Karen's recorder filled the room with soft and pleasant music. Early in the evening, Mike noticed Pat's hand and the many diamonds. As the night progressed the tempo of the play increased. Finally, a lull in the activities occurred. Taking advantage of this, Jerry pondered the arrangement in front of him. Mike interrupted Jerry's reverie and said, "Let's hear the score." They listened carefully and commented on their performance. When the comments were all heard, exhausted but happy, Karen's friends went home.

Source: Anderson, R. C. 1977. The notion of schemata and the educational enterprise. In *Schooling and the acquisition of knowledge*, ed. R. C. Anderson, R. J. Spiro, and W. E. Montague, pp. 415–431. Hillsdale, NJ: Lawrence Erlbaum Associates.

Sample Passage #4

Rocky slowly got up from the mat, planning his escape. He hesitated a moment and thought. Things were not going well. What bothered him most was being held, especially since the charge against him had been weak. He considered his present situation. The lock that held him was strong but he thought he could break it. He knew, however, that his timing would have to be perfect. Rocky was aware that it was because of his early roughness that he had been penalized so severely—much too severely from his point of view. The situation was becoming frustrating; the pressure had been grinding on him for too long. He was being ridden unmercifully. Rocky was getting angry now. He felt that he was ready to make his move. He knew that his success or failure would depend on what he did in the next few seconds.

Source: Anderson, R. C., R. J. Spiro, and W. E. Montague, eds. 1977. *Schooling and the acquisition of knowledge.* Hillsdale, NJ: Lawrence Erlbaum Associates.

Sample Passage #5

With hocked gems financing him, our hero bravely defied all scornful laughter that tried to prevent his scheme. "Your eyes deceive," he had said. "An egg, not a table, correctly typifies this unexplored planet." Now three sturdy sisters sought proof. Forging along, sometimes through calm vastness, yet more often very turbulent peaks and valleys, days became weeks as many doubters spread fearful rumors about the edge. At last from nowhere welcome winged creatures appeared signifying momentous success.

Source: Dooling, D. J., and R. E. Christiansen. 1977. Episodic and semantic aspects of memory for prose. *Journal of Experimental Psychology: Human Learning and Memory* 3: 428–436.

Debriefing

When Working With Teachers

Use the NSES quote to contrast instruction that is merely hands-on to that which is truly minds-on science teaching (MOST) and catalyzes the "most" learning. Reading about science is a necessary but insufficient component of effective science curriculum-instruction-assessment (CIA). Having students critically read textbooks is a wonderful supplement to (i.e., confirm, reinforce, enrich, and extend), rather than substitute for, doing science with physical phenomena and/or interactive multimedia simulations. Intelligent CIA seeks to uncover the big ideas and nature of science and develop scientific habits of mind rather than merely cover the textbook. The reciprocal, synergistic relationship between science-inquiry skills and reading-process skills was noted more than 30 years ago in several NSTA publications:

- The chapter titled "Science: A Basic for Language and Reading Development" in *What Research Says to the Science Teacher, Vol. 1* (Rowe 1978)
- The article titled "Science and Reading: A Basic Duo" (Carter and Simpson 1978) in an issue of *The Science Teacher*. A comparative

analysis of corresponding skills in science and reading suggests that science and reading reinforce each other if taught properly (see Internet Connections: Just Read Now).

When Working With Students

Use this discrepant-event reading activity and the quotes from Francis Bacon and Dr. Seuss to discuss the shared, minds-on, theory-driven yet empirically grounded nature of reading and science. Also, consider sharing the longer quote about how much time practicing scientists spend reading (Introduction, p. xvii). Both reading and science are interactive, constructive processes in which what the learner already knows filters and colors in a top-down, schema-driven fashion what is processed (from either a text passage or an experiment) in a bottom-up, experience-driven fashion. Efficiently learning science from reading text requires both pre- and postreading experiences with live and/or virtual phenomena such as inquiry-oriented demonstrations, hands-on explorations, computer simulations, and multimedia. These types of concrete experiences activate, challenge, and extend students' prior conceptions while simultaneously generating a "need to know" motivation that can be satisfied by further reading, research, and reflection. Appropriately selected and critically analyzed visuals (i.e., "a picture is worth a thousand words"; see Extension #1) and analogies are also powerful big-picture, context-setting instructional aids. Students need to see how they can use these varied types of science learning experiences to help them make meaning by constructing interconnected conceptual networks. If science teachers teach their students how to learn to read so they can read to learn from science textbooks, then reading assignments will not be another source of stress (or something to skip), but rather a source of strategic, "just in time" support and scaffolding for their learning. Reading can be assigned to students in any phase of the 5E Teaching Cycle (see next section), though conventional textbooks are probably most useful in the Explain or Elaborate phases.

Teachers can periodically model for their students how to use active, metacognitive strategies before, during, and after they read the textbook. The content-area reading books and Internet Connections listed in Activity #4 contain information on this topic. One instructional strategy is for the teacher to explicitly demonstrate a live "interrogation" of a sample reading assignment (e.g., SQ3R: Survey/Skim + Question, + Reads + Recites/Recalls/Retells, and/or Records + Reviews). Read-aloud and think-aloud modeling of the internal, conceptual "dissection and reconstruction" process can be visually traced using a concept map (Good, Novak, and Wandersee 1990) or some other type of graphic organizer (see Internet Connections). After modeling, teachers can use reciprocal teaching and guided practice with feedback to coach readers (and teach them how to coach each other) how to use this process themselves (see the websites of Learning Point Associates [p. 75] and Reading Quest.org and reciprocal teaching [p. 61]). Emphasize that mindlessly memorizing isolated science-sounding words without placing them in an overall context where "one thing leads to another" to tell an interconnected "science story that makes sense" is not learning. Teaching students how to read to learn science is a yearlong responsibility of every science teacher. Students need to understand that science textbooks offer unique cognitive challenges even for strong readers.

Extensions

1. *Fossil Footprints: Reading Clues and Science Storytelling:* Activity #5, "Proposing Explanations for Fossil Footprints" (ch. 6, pp. 87–89) in the book *Teaching About Evolution and the Nature of Science* (National Academy of Sciences 1998) adapts the 1960s Earth Science Curriculum Project inquiry activity into a mini 5E Teaching Cycle. This wordless picture puzzle can be downloaded from the Internet Connections. (*Note:* The third bullet in the Internet Connections also includes the Mystery Tube and Dinosaur Bones activities.) Students develop one or more plausible explanatory stories based on their observations of and inferences about three successive panels of animal footprints. The pattern of footprints could suggest two animals that vary in size, age, gender, species, and/or mode of transportation (e.g., ground and/or flight). Plausible stories include two species with one larger predator and one smaller prey; the same species with mother and offspring interacting or male and female engaged in copulation or fighting over the same food source (e.g., seeds on the ground); or the same or different species in the same location at different

times scavenging the ground for food. Assuming the tracks were made at the same time, in the final panel one organism could be on top of, inside (having been eaten), or have flown away from the other.

For an additional challenge, the three panels of this picture puzzle can be given as separate, nonsequenced strips, and students can be asked to place the three in a sensible time-ordered sequence. The same observations allow for rich and varied inferences; exposure to additional data provided in each successive panel and questioning of starting assumptions also alter previous interpretations. This classic activity models how scientists use data to tell sensory-based, sensible, skeptically studied "stories" that remain tentative and open to further revision in light of new evidence. It links meaning-making in science and reading via a common emphasis on storytelling in light of empirical evidence, logical argument, and skeptical review.

2. *Tricky Text and Testing Trivia*: Select an ambiguous passage that has not been previously discussed, and challenge groups of teachers to construct short tests on the passage that consists of low-level Bloom's taxonomy recall-type assessment items (i.e., true-false, fill-in-the-blank, and multiple-choice) for their test-savvy teacher-peers to take. Alternatively, more difficult tests can be designed and comparative analysis done on the results for groups of teachers who were given the "big picture" context for the passage compared to those who were not. In either case, after different groups take the tests developed by their peers, discuss how teachers can help their students learn how to use their textbooks more effectively as learning aids, and design integrated and iterative curriculum-instruction-assessment (CIA) that develops meaningful, memorable understandings as a result of students seeing and doing "real science" in conjunction with appropriately timed reading about and reflecting on science texts. See also *More Brain-Powered Science*, Activity #21, Extension #1, for additional examples of "tricky text and testing trivia."

3. *Science Stories: Strangers in a Strange Land and Crossing Cultural Contexts*: Science teachers have commonly lived inside the borders of the land of science for so long that they forget how "foreign"

its language, customs, and mores are to relatively uninformed student-outsiders. To get a better sense of the challenges student face in "border crossing," it is useful for teachers to brainstorm and discuss how reading a science textbook is quite different from reading everyday materials such as novels, short stories, and newspaper articles. The Answers to Questions in Extensions includes a table that depicts some of these differences.

Other activities that teachers can use to help students understand the shared inquiry nature of science and reading include the "Blind Men and the Elephant" parable poem (see Internet Connection: Biology Corner and *More Brain-Powered Science*, Activity #9, Extension #2), and any of Sir Arthur Conan Doyle's original Sherlock Holmes stories or the *Sherlock Holmes* movie (2009) or BBC *Masterpiece Mystery!*/PBS *Sherlock* series (2010), which feature the inquiry skills of this master sleuth. Both the movie and TV versions have segments where his high-speed thinking processes are cinematically slowed down so the viewer can see and skeptically review the logical connections he makes between empirical observations (or perceptions) and inferences (or conceptions). Though real scientific insight and discoveries rarely happen so quickly and scientists don't need to be somewhat disturbed geniuses, the stories and movies draw the reader or viewer in to the excitement of scientific processes in action. In a similar vein, the CBS TV series *Numb3rs* features a brilliant mathematician and his mentor, a physicist, using their scientific-reasoning skills (often explained with visual analogical models) to help solve FBI investigations.

4. *The Two Cultures Problem Revisited*. In 1959, C. P. Snow warned against the widening cultural gap between the sciences (including mathematics) and the humanities. Teachers can examine this classic book (see Internet Connections) through the lens of today's debates over scientific literacy and the "border crossings and intercultural exchanges" between science, ELA/reading, and/or social studies teachers in their schools and districts. Collaboration could create win-win synergy that benefits both groups of teachers and especially their students. See Internet Connections: Science News on the Internet for sources of articles

that might be used in interdisciplinary activities and assignments. Also, develop a 2 × 2 matrix table with an *x*-axis that runs from low to high content knowledge and a *y*-axis that runs from low to high reading skills and discuss what students in the four different quadrants need in terms of instructional support from science and ELA teachers. Are there any other student variables that are relevant to the interaction of science and reading? See Answers to Questions in Extensions.

5. *Ambiguous Figure Optical Illusions* (Activity #5 in *Brain-Powered Science*) can be considered analogous to ambiguous text because in both cases prior knowledge and beliefs and the specific context of the presentation influence what we see and the stories we tell about reality.

Internet Connections

See those listed for Activities #3 and #4, as well as the following:

• Ambiguous Figure Optical Illusions: *www.optillusions.com* and *http://planetperplex.com/en/index.html*

• Biology Corner:

Blind Men and the Elephant Poem: *www.biologycorner.com/worksheets/elephant_poem.html*

Identify controls and variables in five one-paragraph-long cartoon stories: *www.biologycorner.com/worksheets/controls.html*

Scientific Method Stories (read and answer questions on beriberi and penicillin): *www.biologycorner.com/worksheets/scientificmethodstories.html*

• Fossil Footprints Activity:

http://books.nap.edu/openbook.php?record_id=5787&page=87

http://orsted.nap.edu/openbook/0309063647/gifmid/89.gif (actual picture puzzle)

See also *www.teacherlink.org/content/science/class_examples/Bflypages/timlinepages/nosactivities.htm#footprints; www.teacherlink.org/content/science/class_examples/Bflypages/timlinepages/nosactivities.htm#mystery;* and *www.teacherlink.*

org/content/science/class_examples/Bflypages/timelinepages/ nosactivities.htm#bones.

- Graphic Organizers:

    Freeology (free download PDF files): *http://freeology.com/ graphicorgs*

    Houghton Mifflin Harcourt Education Place (free, downloadable PDF files): *www.eduplace.com/graphicorganizer*

    Inspiration: *www.conceptmapping.com/productinfo/inspiration/ index.cfm*

    Institute for Human and Machine Cognition (free concept mapping software and research): *http://cmap.ihmc.us*

    Learning Point Associates: *www.ncrel.org/sdrs/areas/issues/ students/learning/lr1grorg.htm*

    Periodic Table of Visualization Techniques (creatively displays more than 100 types): *www.visual-literacy.org/periodic_table/ periodic_table.html*

    Wikipedia: *http://en.wikipedia.org/wiki/Graphic_organizer*

- Just Read Now: Science: *www.justreadnow.com/content/science/ index.htm*

- Memory: Myths, Mysteries and Realities (a synopsis of research): *www.answers.com/topic/memory-myths-mysteries-and-realities*

- Science News on the Internet (reputable sites by legitimate scientific sources):

    BBC Science and Nature: *www.bbc.co.uk/sn*

    CNN: *www.cnn.com/TECH/science/archive/index.html*

    EurekAlert!: *www.eurekalert.org* (sponsored by AAAS)

    Science *Daily*: *www.sciencedaily.com*

    The New York Times: *www.nytimes.com/pages/science/index.html*

- *The Two Cultures and the Scientific Revolution* (1951 lecture and book by C.P. Snow): *http://sciencepolicy.colorado.edu/students/ envs_5110/snow_1959.pdf*

    See also synopsis at *http://en.wikipedia.org/wiki/ The_Two_Cultures*

Answers to Questions in Procedure, steps #2 and #3

The lack of a title, visual image, or other unambiguous context clues in the passages makes it difficult for readers to make sense of the meaning even though they most likely have the relevant prior knowledge about the underlying phenomenon. Reading even well-written science text is likely to be more difficult for students than this passage because the science text may include many "foreign" words that students do not know or "everyday" words that have unique meanings in science (i.e., *work*, *energy*, and *force*); students may lack sufficient prior experiences with the phenomena; understanding may involve abstract science concepts or mathematical skills that have not been adequately developed prior to the reading assignment; and/or overly complex, overly simplified, or otherwise poor-quality or uncritically examined visuals can distract and confuse readers and create misconceptions. (*Note:* Activity #13 has students explore a case where textbook visuals "lie" when they don't need to.)

The "correct" contexts or titles for the five sample reading passages are:

1. "Making and Flying a Kite"

2. "Doing Laundry" or "Washing Clothes" (or perhaps dishes)

3. "Playing Cards" (or perhaps a Musical Quartet); the original title was "An Evening at Play"

4. "The Prisoner" and "The Wrestler" are alternative original titles.

5. "Columbus's Voyage to the New World." The three sturdy sisters are the Nina, Santa Maria, and Pinta, the names of Columbus's three ships.

Answers to Questions in Extensions, #3 and #4

Extension #3: Contrasting the Science Textbook Style Writing and Everyday Reading Materials

Science Textbooks	Everyday Reading
nonfiction; expository writing with passive, "hidden" author's voice; and few, if any, human characters are evident	fictional or nonfictional narrative stories told in an active voice and human characters and their interactions are critical
specialized terminology (Greek and Latin parts) and uses common words in restricted ways	relatively few specialized terms or unique or restricted definitions
linguistically and conceptually dense or "supersaturated" and require multiple readings (and other experiences) to be understood	typically can be easily skimmed for the gist of the story without additional experiences needed for understanding
didactic, pedantic tone and informational text that require readers to have a facility with empirical evidence, logical arguments, and skeptical review to reconstruct the author's meaning	conversational tone and story text make it relatively easy for the reader to follow along and reconstruct the author's meaning
"foreign" words for "foreign" things that lie outside of everyday experiences and perhaps unaided sensory capabilities (i.e., abstract)	easily relates everyday life (i.e., concrete) and the personal experiences, emotions, interests, and prior knowledge of the readers
nonpictorial, representational visuals (e.g., cross section, cutaways, explosion diagrams and sketches, charts, tables and off-scale models) that can complement or compete with the text for the readers' attention (see Mayer 2009a)	photos or realistic sketches require little interpretation; typically they can even be ignored if desired as they are simply optional "pretty pictures"
often demands mathematical literacy (numeracy). Science is a hybrid, multimodal language that integrates unique theory-dependent words, mathematical symbolisms, and representational-type visuals	typically requires little to no mathematical understanding or theoretical ways of "seeing" to make sense of the text
only allow naturalistic, agnostic explanations that may seem restricted or even in conflict with deeply held beliefs of some readers	may include references to supernatural entities—such as disembodied spirits, an afterlife, and God—that create relevant connections for some readers

Extension #4

The following 2 × 2 matrix depicts the interaction of reading skills and science content knowledge. Other relevant variables include a student's motivation to learn in general, content-specific interests, work ethic, and study habits. A student's zone of proximal development (ZPD) and the quantity, timing, and nature of the instructional support he or she needs depends on the interaction between a number of variables. The 5E Teaching Cycle approach that is modeled in the next section of this book is a research-informed approach to providing "different strokes for different folks" across a unit of study.

	High	Science novice but academically oriented (needs scaffolded experiences with science phenomena linked to reading passages)	Independent reader-learner (capable of more self-selected, discovery-oriented reading and experimentation)
Reading Skill	**Low**	Frustrated learner headed for failure (needs CIA that synergistically develops strength in both areas of deficiency while motivating effort and self-efficacy)	Potential future scientist (like all the others) (needs level-appropriate but motivational science text as a lever to develop reading skills and full scientific literacy)
		Low	**High**
		Content Knowledge	

Section 3:
Integrated Instructional
Mini Units:
5E Teaching Cycles

Resurrection Plant
Making Science Come Alive!

You really need to take better care of yourself.

Expected Outcome

An inanimate, seemingly dead, 3–5 in. ball-shape brown object placed in water and exposed to sunlight is observed to "come back alive" and turn into a vibrant green photosynthetic plant over a period of anywhere from several hours to one day. When removed from water, the object gradually returns to its dried up, ball-shaped, dead-looking appearance. This cycle is repeatable.

Science Concepts

The resurrection plant provides an engaging, discrepant introduction to concepts such as the characteristics of life, water as an essential requirement of life, photosynthesis, plant adaptations, and species evolution and survival of the fittest to specific environments (even "extreme" ones). The name "resurrection plant" refers to any of a variety of species of ancient fernlike plants that are native to dry, desert, or semi-arid regions that have the biological adaptation to survive in a sort of "plant hibernation or suspended animation" for years without water but return to their full greenery within hours of being placed in water. These species have hygroscopic qualities that enable them to curl up and greatly reduce their metabolic activity when dry and to unfold and "come back to life" when adequate moisture is present. The resurrection plant's water/environment relationship is a *system* where *constancy* and *change* are evident (see Appendix C for other "big ideas" in science). All organisms have homeostatic mechanisms to enable them to maintain a dynamic equilibrium with their environments. If the environment changes dramatically, organisms must move, die, or, in the case of some species, enter into a state of relative physiological dormancy until conditions become more favorable.

Science Education Concepts

Investigations of this discrepant phenomenon can be used to quickly demonstrate to teachers the idea of a 5E Teaching Cycle (Engage, Explore, Explain, Elaborate, Evaluate) as an instructional model for multiday to unit-level planning. It also demonstrates how to use discrepant events to activate, diagnostically assess, and challenge students' prior conceptions. Finally, the activity can be used as a *visual participatory analogy* for how seemingly "dead" students can "blossom and grow" (or not) depending on how well the curriculum-instruction-assessment (CIA) learning environment aligns with their interests, needs, and capabilities.

Teaching can be viewed analogically as a type of gardening that involves "planting seeds in fertile soil" (i.e., placing students in CIA environments that contain learner-appropriate, sensory, and conceptual "grounding"); "pulling weeds" (i.e., finding and removing misconceptions that restrict the growth of new ideas); and "applying feed" (i.e., providing the proper amounts and schedule of "smart" CIA interventions that serve not as preformed and predigested food, bur rather as fertilizers to catalyze growth). In this context, learning can be viewed as the natural, evolution-driven capability of students (seeds) to grow into vibrant plants that can produce their own food and transform their environments. When placed in appropriate environments, students mature into self-directed learners who construct improved conceptual networks and use both speculative/creative and skeptical/critical reasoning to identify and solve problems on their own. Activity #1 in *Brain-Powered Science* (O'Brien 2010) develops this and other analogies related to the dynamics of the "ecological system" of teaching-learning.

Materials

- If the activity is done as a teacher demonstration-experiment, at least two resurrection plants are needed. If used as a student hands-on exploration, every group of two or three students will need one or two plants.

  Resurrection plants are available from local greenhouses and the following vendors:
  - Carolina Biological (*www.carolina.com*; 800-334-5551): Resurrection Plant: *Selaginella lepydophyll* (#157010 for $5.25)
  - *Pow! Science Store* (*www.powscience.com/store*; 888-708-2019): Amazing Dinosaur Plant: *Selaginella lepydophyll* ($6.99)
  - *Ward's Natural Science* (*http://wardsci.com*; 800-962-2660): Resurrection Plant: *Selaginella lepidophylla* (#86 V 5420 for $6.75)

Note: Be sure the plants are completely dry (to avoid mold formation) before returning them to plastic bags for storage and later reuse for years of repeatable cycles of de- and rehydration.

- Hand lenses or 30× handheld microscopes are useful if available.

Points to Ponder

The observer listens to nature; the experimenter questions and forces her to reveal herself.

—Georges Cuvier, French zoologist, anatomist, paleontologist, and anti-evolutionist (1769–1832)

The survival of the fittest, which I have here sought to express in mechanical terms, is that which Mr. Darwin has called "natural selection, or the preservation of favored races in the struggle for life."

—Herbert Spencer, English philosopher and psychologist (1820–1903)

The challenge of all of us who want to improve education is to create an educational system that exploits the natural curiosity of children, so that they maintain their motivation for learning not only through their school years but throughout their lives.

—*Inquiry and the National Science Education Standards* (National Research Council 2000)

Procedure

This hands-on exploration is designed as a mini 5E cycle. If only one or two plants are available, it can be done as a teacher demonstration by using a document camera to project contrasting images of a brown, "dead" plant versus a green, living one (the latter, whose roots have been pre-soaked in water and exposed to sunlight for one or two days before the demonstration).

1. *Engage:* Challenge learners working in groups of two or three to examine the object provided to try to come to a group consensus as to whether it is living or dead, or if it never was living (use hand lenses or handheld microscopes if available). Questions to consider include the following:

   a. What are the distinguishing characteristics of living things?

b. Why might it be difficult to determine if something is alive or not?

c. What environmental factors are essential for living organisms to prosper?

d. How do we distinguish plants from animals?

e. If you were to decide the specimen was once living but is not now, would this be an observation or an inference? Why? List as many observations as possible (at least three) that support this belief.

f. What role does prior knowledge play in your determinations? Are our prior conceptions always valid, or can they sometimes lead to incorrect predictions?

2. *Explore:* Share the Points to Ponder quote from Georges Cuvier to contrast simple, casual observations from more focused, inquiry-driven explorations. Challenge the learners to design and carry out one or more simple experiments that could help support or refute the "once living, now dead" hypothesis. *Note:* Teachers may wish to have set a specimen in water a day or two before this activity. If students seem lost, tell them that the green living plant (patted dry and displayed separately from its cup of water) you showed them previously looked like their "dead" one. Ask them to guess what environmental variable(s) you changed. Students will likely consider sunlight, water, and nutrients. Challenge students to develop controlled, quantitative tests such as weighing the dry versus the watered plants; measuring the volume and mass of water added versus that which seemingly disappears (i.e., lost by evaporation, used during photosynthesis, or taken up by the plant and retained or transpired); and considering how to control for evaporation and transpiration. Place their specimens in water in the presence of sunlight or artificial light, and have students re-examine the specimens the next day.

If desired, time-lapse photography (search YouTube or TeacherTube, or learn how to make your own at the Plants in Motion website) can be used to show the two-hour or longer "resurrection" process in an accelerated time frame of approximately one minute. Plants have internally driven growth and/or orientation movements (or tropisms) in response to environmental stimuli such as light and gravity. However, plant movements typically happen at a much slower rate and extent than animal movements.

3. *Explain*: In science, experimental answers to even such deceptively simple questions as "Is it living or nonliving?" provide the basis for the next round of questions and explorations. In the case of the resurrection plant, when put in water, the plant's metabolic rate (i.e., both photosynthesis and respiration) increases and the plant takes on the characteristic green color of photosynthetic, autotrophic plants. It is indeed alive! As a class, discuss the individual biological survival and species reproductive value of this unusual "resurrection" behavior in a species that dates back some 290 million years. Interactive teacher presentations (i.e., including the use of multimedia resources and simulations) and targeted reading assignments from a science textbook, the internet (see below), and other sources can be used to formally introduce and develop concepts such as the characteristics and requirements for life, species-environment fitness, biological adaptations, and other evolution-related concepts. Share the Herbert Spencer quote about survival of the fittest and clarify the distinction between "best" and "optimized for a particular set of environmental conditions." Ask: What are the biological advantages of a plant "pulling up its roots" and becoming ball-shaped? Think about the "extreme" environmental conditions of the locations featured in western movies, where tumbleweeds are commonly featured in deserted towns. If this adaptation is a good idea, why don't all plants have the ability to become mobile tumbleweeds?

4. *Elaborate:* Depending on the age of the students and their prior experiences with designing semi-independent experiments, they can try to discover the optimal lighting (i.e., intensity and color range) and/or temperature conditions for the fastest regeneration time; microscopic investigations of the vascular system in the dry (i.e., approximately 3% of normal water) versus hydrated state; the net uptake of carbon dioxide and release of oxygen; and so on. Students can do library and internet research to discover more about the different species that exhibit the unusual "resurrection" behavior and any current uses and research studies related to this species-environment "system fit." Also, any of the Extensions for

[handwritten margin notes: define adaptation]

[handwritten margin notes: Why might the resurrection plant have this adaptation?]

[handwritten margin notes: resurrection plants are optimized for (tumbleweeds) dry conditions]

this activity or the following historical case study could be used during this phase (*www.ActionBioscience.org* has articles that discuss how to use historical case studies in teaching).

Van Helmont's Historic "Miss-take": The resurrection plant can be linked to the classic willow tree experiment of Flemish physician and alchemist Jan Baptist van Helmont (1579–1644; see Internet Connections, Videos and DVD: Teachers' Domain: *Photosynthesis* and Wikipedia). Working at a time when the ideas of controlled experiments and quantitative measurements were just starting to emerge, van Helmont tackled the question of the source of the mass gain in plants. Over a five-year period, his willow tree gained 164 lbs. while the soil it was planted in lost only 2 oz. He concluded (incorrectly) that the mass gain was primarily due to the water he had added. Based on their prior experiences with plants, most students will probably harbor similar misconceptions that the growth in size and mass of a plant as it develops from a seed to a full-size plant is due to the soil and/or the water. Because the resurrection plant appears to develop in the absence of soil, the water hypothesis may seem more likely.

Challenge students to account for the mass gain from a seed to a cut, *dry* piece of a tree. The fact that most of the mass comes from carbon dioxide in the atmosphere is astonishing to most students even if they have memorized the photosynthetic reaction (see the Debriefing discussion). The chemical reaction shows that solar-powered photosynthesis of carbon dioxide and water produces glucose that plants can burn for fuel or convert to more complex sugars or polymers such as starch and cellulose for building its physical structures. (*Note:* The reaction also produces oxygen that is released to the environment.) Ironically, van Helmont was one of the first experimenters to demonstrate that there were different kinds of "gases." In fact, he coined the word *gas* from the Greek word *cháos* and identified what we now call carbon dioxide, carbon monoxide, nitrous oxide, and methane. Scientific understanding of the chemistry of the photosynthetic process was not elucidated until the Nobel Prize–winning work of Melvin Calvin in the 1950s (see Wikipedia).

5. *Evaluate:* Learners can be asked to

- write essays that explain how the resurrection plant exhibits the characteristics of life and challenges one or more commonly perceived distinctions between plants and animals (i.e., mobility and limited, imperceptibly slow responses to the environment);
- prepare a research report on other plants (and/or their seeds) that move from place to place or animals (e.g., coral) that are basically stationary for most of their life;
- prepare PowerPoint presentations on the long-term, "extreme environment" survival ability of yeast, various plant seeds (e.g., corn in Egyptian tombs), spores, and viruses that feature the theme of the characteristics of and requirements for life; or
- plan, perform, and present results of an experiment with other organisms (see the Extensions for this activity or the Activity #7 Extensions).

Debriefing

When Working With Teachers

Discuss the importance of the surprise factor in challenging students' prior limited conceptions of nature's ingenuity in general and the creative adaptability of plants in particular, and the analogy between this activity and the idea that classroom environments that feature high-quality CIA allow students to "blossom and grow" (or the lack thereof can cause them to "shrivel up and stop growing"). On a related note, consider how the idea of biological survival of the fittest is an apt analogy for the "evolution" of scientific ideas across history and the changing "conceptual ecology" of individuals' understanding of science within and across the K–12 school years. Also, see the Internet Connections: BSCS and the Miami Museum of Science for more information on the 5E Teaching Cycle as a CIA model that helps scaffold students' growth in understanding science.

A most engaging teacher and student resource book on the science of life is *The Way Life Works: The Science Lover's Illustrated Guide to How Life Grows, Develops, Reproduces, and Gets Along,* by Mahlon Hoagland and Bert Dodson (1998). The book features full-color cartoonlike drawings and visual analogies and seven major themes related to the characteristics of life. Reading assignments from such well-designed books can complement discrepant phenomena to exploit adolescents' natural curiosity. See also Hackney and Wandersee (2002) and Harrison and Coll (2008) for other biological analogies; the former contains creative student worksheets, and the latter has teacher examples and research.

The Annenberg Foundation's *The Private Universe: Minds of Our Own*, Program 2: *Lessons From Thin Air* (1 hr.; see Internet Connections) is a free professional development video-on-demand that depicts Harvard graduates in caps and gowns being asked to account for the primary source of mass gain from seed to (dry) tree. None suggest carbon dioxide as the answer. The video further explores how this important idea can be developed at the middle school science level. The idea of activating and challenging students' prior (mis)conceptions is featured (e.g., a discrepant-event demonstration with dry ice or solid carbon dioxide that shows gases have mass). See also plant nutrition, photosynthesis, and gas exchange in Driver et al. 1994 (pp. 30–34) and Extensions #1 and #2 for this activity.

When Working With Students

In addition to the science content, discuss the complementary roles of observational, experimental, and theoretical biology. Life science classrooms should contain living organisms beyond the students and unseen microbes. Growing, maintaining, and studying a variety of both commonplace and unusual plants allow students to grasp the importance of plants and develop a respect for the diversity and interdependence of life. Inviting students to respectfully explore living organisms both inside the classroom and outside in natural settings is an effective means of counteracting "nature deficit disorder" (Louv 2008; see also Internet Connections: Wikipedia). Breaking out of the limitations of "2 × 4 teaching" (i.e., instruction constrained by the two covers of the textbooks and four walls of the classroom) suggests an expanded conception of laboratory learning (see Internet Connections: National Association of Biology Teachers: The Role of the Laboratory and Field Instruction).

Extensions

1. *Playing With Perplexing Plants:* Demonstrations, hands-on explorations, field trips, video clips, and the internet can be used to

study other plants with somewhat unusual evolutionary adaptations. Plants that are easy to obtain include carnivorous plants (e.g., Venus flytrap), *Mimosa* sensitive plants (sometimes marketed in lab kits under the name TickleMePlant), air fern, duckweed (i.e., the smallest flowering plant), and *Brassica rapa* (used in school labs for their rapid life cycle; see Internet Connections: Wisconsin Fast Plants). *Elodea*, a genus of aquatic plant that is available from aquarium fish stores, is widely used to study photosynthesis and respiration in plants. The typical setup for this experiment uses water with drops of either bromothymol blue (changes color [pH] from yellow [6.0] to green [6.4] to blue [7.6]) or phenol red (yellow color at a pH of 6.4 or lower and a red color at a pH of 8.2 or above) added to it and carbon dioxide bubbled through it. (See the Internet Connections: Elodea Experiments sites for procedures and safety precautions; review the MSDS for hazardous chemicals with students.) All of these unusual types of plants challenge students' misconceptions about nature always fitting into nice, neat, clearly defined boxes. See also Activity #9 for another example of a unique, discrepant-event plant adaptation (i.e., burdock). Additionally, certain parasitic plants may even lack chlorophyll and the ability to photosynthesize their own food (see Internet Connections: Botanical Society of America and Wikipedia).

2. *Multimedia, Misconceptions, and Myopia About Botany:* Most students view plants as inanimate, barely alive, and hardly worth noticing. This plant myopia persists despite the fact that plants are essential to nearly all life on Earth (with the exception of chemosynthetic bacteria) and serve as sources of food, fiber (clothing), pharmaceuticals, and building materials for humans. Though science education organizations such as NABT and NSTA have position statements about the classroom use of animals, they do not have similar statements about the use of plants. Articles on *www.Actionbioscience.org* and in resources provided by the American Society of Plant Biologists challenge "animal chauvinism" and argue for the full inclusion of plants in K–12 classrooms (see Internet Connections). Also, the videotape and DVD programs listed on page 94 use time-lapse and macrophotographic techniques to portray plants as exciting, "sexy," and

even a little macabre. These films challenge many misconceptions that students have about plants.

Some botanical misconceptions arise from misuse of scientific terminology in the popular culture. For instance, Miracle-Gro, Jobe, and other manufacturers sell "plant food spikes" despite the fact that plants are autotrophs! These products are fertilizers that provide essential elements such as nitrogen, phosphorus, and potassium as well as trace elements; none of these become part of the sugars or starches that the plants produce as food via photosynthesis. Other misconceptions come from poor instruction, such as the failure of many textbooks to sufficiently emphasize that plants respire as well as photosynthesize and that they have co-evolved with and use animals as much as animals rely on plants. Still other misconceptions relate to the discrepancy that some plant species are endangered and need to be conserved, while other invasive, exotic species need to be controlled and removed. See *www.Actionbioscience.org* for 50 misconceptions to avoid when teaching about plants, and *More Brain-Powered Science*, Activity #22, Diagnostic Assessment (Test #2, Rooting for Plants) for a quick pre-instructional test that highlights a number of plant misconceptions.

3. *Triops: Animals Can "Resurrect" With Water Too! Triops* is a genus of small aquatic crustaceans that look like miniature horseshoe crabs and also date back to the time of dinosaurs (i.e., 180 million years ago to the Jurassic period). Like some other arthropods, *Triops* can experience a delay in development in response to regularly occurring adverse environmental conditions such as temperature extremes, drought, or reduced food supplies. Called diapause, this physiological state of dormancy is a biological adaptation that increases survival rates. The dry tiny eggs can be purchased in kits and can remain in their dormant state for years, but they will grow to an adult size of 1–3 inches in less than 30 days when placed in water. See Internet Connections: *Triops*. See also Activity #7, Extension #2, for a smaller species of brine shrimp that displays similar behavior.

Internet Connections

- American Society of Plant Biologists: Education (resources): *www.aspb.org/education*

- Articles and Resources on Plant Misconceptions and the Lack of Curriculum:

    ActionBioscience.org: *www.actionbioscience.org/education*. See Classroom Methodology: Using Case Studies to Teach Science (by Clyde Freeman Herreid) and Tapping Into the Pulse of the History of Science With Case Studies (by Douglas Allchin); *www.actionbioscience.org/education/hershey.html#primer* and *www.actionbioscience.org/education/hershey3.html*

- Biology Corner: Plants (labs and simulations): *www.biologycorner.com/lesson-plans/plants*

    Water in Living Things (lab): *www.biologycorner.com/lesson-plans/scientific-method*

- Botanical Society of America:

    Carnivorous Plants: *www.botany.org/Carnivorous_Plants*

    Parasitic Plants: *www.botany.org/parasitic_plants*

    Statement on Evolution: *www.botany.org/outreach/evolution.php*

- *BSCS 5E Instructional Model: Origins, Effectiveness and Applications:*

    www.bscs.org/pdf/5EFull Report.pdf (65 pages)

    http://bscs.org/pdf/bscs5eexecsummary.pdf (19 pages)

- Center for Plant Conservation: *www.centerforplantconservation.org/welcome.asp*

- Elodea Experiments (photosynthesis and respiration):

    www.accessexcellence.org/AE/AEC/AEF/1996/linhares_lab.php (snail and Elodea lab)

    http://cibt.bio.cornell.edu/labs/eeb.las (choose Photosynthesis and Respiration in Elodea; also see Plant Game)

    www.biologyjunction.com/carbon_dioxide_use_in_plants.htm (uses phenol red)

*www.keystonecurriculum.org/2008middleschool/LESSONS/S_
TerrestrialSequestration.doc*

- Miami Museum of Science: Constructivism and 5E: *www.
miamisci.org/ph/lpintro5e.html*

- Missouri Botanical Garden: Charms of Duckweed: *http://mobot.
org/jwcross/duckweed/duckweed.htm*

- National Association of Biology Teachers: Two NABT Position
Statements:

   Teaching Evolution and Role of the Laboratory and Field
   Instruction in Biology Education: *www.nabt.org/websites/
   institution/index.php?p=35*

- National Center for Science Education: Defending the Teaching
of Evolution in Public Schools: *http://ncse.com*

- National Science Teachers Association Position Statements:

   The Teaching of Evolution: *www.nsta.org/about/positions/
   evolution.aspx*

- *Private Universe: Minds of Our Own*: Program 2. *Lessons From Thin
Air:* Video-On-Demand: *www.learner.org/resources/series26.html*
(free VOD)

- Resurrection Plant: *http://faculty.ucc.edu/biology-ombrello/POW/
resurrection_plant.htm*

- Science and Plants for Schools:

   Teaching resources: *http://www-saps.plantsci.cam.ac.uk/index.htm*

   Xerophytic adaptation: *http://www-saps.plantsci.cam.ac.uk/
   records/rec254.htm*

- Triops: *http://en.wikipedia.org/wiki/Triops*

   www.youtube.com/watch?v=4quPALV24Ww (fact-based slide
   show with music)

   *www.discoverthis.com/triops.html?gclid=CMOBydDc8aMCFYp_5
   Qod3mpy1g* (kit)

   *www.amazon.com/Triops-Inc-DLX-Deluxe-Kit/dp/
   B000A6QMTM* (kit)

- Videos and DVDs on the Unique Adaptations of Plants:

Carolina Biological: *www.carolina.com;* 800-334-5551 (#495932D) *Sexual Encounters of the Floral Kind*: DVD tape (60 min.) for $82.95

David Attenborough's *The Private Life of Plants*: 2 DVDs (300 min.) for $29.37: *http://shop.abc.net.au/browse/product. asp?productid=724292.* This BBC series is described at *http:// en.wikipedia.org/wiki/The_Private_Life_of_Plants*

Disney Educational Productions: *Bill Nye the Science Guy*: *Flowers* and *Plants* ($29.99/26 min. DVD): *http://dep.disney. go.com*

Nature (PBS series): *The Seedy Side of Plants*: *www.pbs.org/ wnet/nature/plants*

Plants-In-Motion (free time-lapse photography QuickTime movies and how to make your own): *http://plantsinmotion.bio. indiana.edu*

Resurrection plant: search Google Videos for a variety of time-lapse images

Teachers' Domain: *Living Life as a Plant* (3 short video clips): *www.teachersdomain.org/resource/lsps07.sci.life.oate.lplifeasplant*

Teachers' Domain: *Photosynthesis* (2:25 min. video clip and background essay): *www.teachersdomain.org/resource/tdc02.sci. life.stru.photosynth*

YouTube.com: online video clips from *The Private Lives of Plants*: Carnivorous Plants

Poisonous Pitcher Plants (4 min.): *www.youtube.com/ watch?v=trWzDlRvv1M*

Venus Flytrap (3 min.): *www.youtube.com/watch?v= ktIGVtKdgwo*

- Wikipedia: *http://en.wikipedia.org/wiki/Main_Page.* Search topics: Jan Baptist van Helmont, Melvin Calvin, duckweed, elodea, invasive species, Mimosa (sensitive plant), nature deficit disorder, parasitic plants, resurrection plant, *Selaginella lepidophylla* (includes 3 hr. time-lapse movie), and tumbleweed:

- Wisconsin Fast Plants: *www.fastplants.org*

Carolina Biological Supply Co., 800-334-5551: *www.carolina. com/category/living+organisms/wisconsin+fast+plants.do*

Answers to Questions in Procedure, steps #1–#4

1. *Engage:*

   a. Living organisms are characterized by their membrane-bound cellular organization, intra- and intercellular movement, ability to metabolize food for energy and building material, excretion of waste, growth and development, ability to maintain homeostasis while responding to stimuli and maintaining a dynamic separation from their environment, asexual and/or sexual reproduction, and evolution over time as a species.

   b. At the macroscopic level at any point in time, living organisms may not necessarily appear to be moving or doing anything to suggest they are alive. At the microscopic level, some level of metabolic activity is always present in organisms that are alive, but some organisms are capable of going into states of relative "suspended animation."

   c. Critical environmental factors include the need for chemical building blocks and energy, water, a certain range of temperatures, oxygen (if aerobic), and so on.

   d. Plants (autotrophs) are multicellular eukaryotes with chloroplasts and cell walls that are able to make their own food through photosynthesis. Animals (heterotrophs) are multicellular eukaryotes that must ingest other organisms as their source of food.

   e. The status of the resurrection plant as once living but now dead would be an inference unless biochemical tests and/or microscopic analyses were done. The former might discover a small amount of green-colored tissue and leaflike structures suggestive of a life function. Microscopic investigations would discover the presence of cells.

   f. Most students expect living plants to be green and rooted in soil and to have flowers—which are characteristics that are not necessarily true for all plants or any particular plant during all portions of its life cycle. Prior conceptions inevitably include misconceptions due to the limited range of our experiences and errors in the actual lessons we learn from any experience.

2. *Explore:* Beyond the simple test of placing the plant in water and noting that in time it turns green, students might be challenged to devise experimental tests such as (a) determining whether the presence, color, or intensity of light seems to affect its recovery time; (b) checking whether the pH or salinity of water the plant is placed in makes a difference; (c) examining the specimen for its cellular structure and intracellular movement; (d) measuring the relative uptake of carbon dioxide and release of oxygen; and (e) exploring the nature of its reproduction.

3. *Explain:* Pulling up its roots and becoming ball-shaped minimizes the plant's exposed surface area, reduces water needs by slowing metabolic activity, and restricts water loss from transpiration at a time when the water supply is inadequate. This unique adaptation also prevents untimely, wasteful release of seeds into an inhospitable environment. As a lightweight ball or tumbleweed, the plant may become mobile via the wind to aid the "search" for water and later seed dispersal. It is important to note that Herbert Spencer's phrase "survival of the fittest" does not mean "best" in any ultimate sense, but rather best suited for a particular set of environmental conditions. For example, in natural environments that never (or rarely) experience water shortages or droughts, the resurrection plant's unique structure and function adaptations would incur a "cost" in terms of matter and energy (and missed opportunities of different investments that could not be made) that would not produce a corresponding benefit and therefore would be maladaptive. For instance, richer, more fertile, naturally irrigated soils provide good permanent locations for plants to "invest in" deeper root systems and taller, woody stems (rather than assume the transient lifestyle of a tumbleweed). Also, it should be obvious that there are size restrictions on this behavior; sizeable plants such as trees are not going to be able to "pull up their roots and roll."

4. *Elaborate:* Internet searches will reveal several species of xerophytic plants that behave like the sample used in this activity:

a. *Selaginellaceae lepidophylla* is a common fernlike, hetereo-sporous, perennial desert tumbleweed plant that ranges from the southwestern United States to El Salvador. A member of the family of spike mosses, this flowerless plant with small, scalelike leaves grows to a height of 10 cm (4 in.). It requires very little water to survive and, when deprived of its minimum requirement, contracts into a ball, breaks free of the soil, and may be blown around by the wind for many years. When the plant senses the presence of adequate water, it sinks its roots and begins to grow again.

b. *Anastatica hierochunticia* (the "Rose of Jericho"), a small, annual herb of the mustard family (Cruciferae), is also commonly called the "resurrection plant." It is a native to the deserts of northern Africa (e.g., Syria) and southwest Asia and blossoms with small white flowers and uses its dried-up, wind-blown tumbleweed state to help seed dispersal.

c. *Polypodium polypodioides* is an epiphyte that occurs in hardwood forests from Delaware to southern Illinois, south to Texas and Florida, throughout tropical America, and then to southern Africa. In the southeastern United States, this resurrection fern is often found on large, spreading branches high up in old live oaks, but it sometimes is also found in cracks in rocks.

Activity 7

Glue Mini-Monster
Wanted Dead or Alive?

Expected Outcome

A drop of clear, colorless, viscous liquid (i.e., a specific brand of modeling glue) assumes the role of an unknown macroscopic, single-celled organism. When placed in a petri dish of water, it is observed to move and interact with other organisms, food, and its aqueous environment in a seemingly lifelike way.

Science Concepts

Characteristics of and environmental requirements for life (e.g., water and food) can be explored in the context of a simulated macroscopic, ecosystem *model* for microscopic organisms and their aqueous environments. Nonliving *systems* without an external source of power cannot display the dynamic equilibrium (or *constancy* amidst *change*) that even the simplest of cellular life forms exhibit. Suggested explorations with yeast provide an opportunity to transition to the microscopic *scale* of actual eukaryotic living organisms.

Science Education Concepts

This sequence of activities can be used to quickly "walk and talk" through the logic of the 5E Teaching Cycle (Engage, Explore, Explain, Elaborate, Evaluate) for teachers. (*Note:* The 5E design is most powerful when planning units of one week or more.) As with Activity #6, this activity serves as a *visual participatory analogy* that models the importance of the learner-to-environment "fit." Students who appear unengaged, inanimate, and mentally "dead" in one classroom can become highly engaged and interactive in science classrooms with a more intelligent curriculum-instruction-assessment (CIA) system in place. Additionally, this activity features the use of fictitious storytelling, historical quotes or vignettes, and discrepant events to engage student interest, elicit their prior knowledge, and establish a "need to know" at the start of a new unit of study.

Materials

- A single tube of Devcon's Duco Cement is covered with opaque tape to conceal its identity
- A petri dish (large diameter preferred)
- Water
- Shavings from a wood pencil and/or pepper (as pseudo-food)
- Overhead projector or document camera
- Blank transparency
- Several packets of baker's yeast
- *Optional:* a toy, battery-powered "smart" car and compound microscopes and/or a microprojection unit

Points to Ponder

In the year 1657 I discovered very small living creatures in rainwater … No more pleasant sight has met my eye than this of so many thousands of living creatures in one small drop of water … I have had several gentlewomen in my house, who were keen on seeing microscopic organisms in vinegar; but some of 'em were so disgusted at the spectacle that they vowed they'd ne'er use vinegar again. But what if one should tell such people in future that there are more animals living in the scum on the teeth in a man's mouth, than there are in a whole kingdom? … All the people living in our United Netherlands are not so many as the living animals that I carry in my own mouth this very day!

—Antonie van Leeuwenhoek, Dutch microscopist (1632–1723)

So nat'ralists observe, a flea

Hath smaller fleas that on him prey,

And these have smaller fleas that bite 'em,

And so proceed ad infinitum.

—Jonathan Swift, Anglo Irish satirist and cleric (1667–1745)

"I say!" murmured Horton. "I've never heard tell
Of a small speck of dust that is able to yell.
So you know what I think? …Why, I think that there must
Be someone on top of that small speck of dust!
Some sort of a creature of very small size,
too small to be seen by an elephant's eyes …

—Dr. Seuss, *Horton Hears a Who* (1954)

Procedure

1. *Engage:* Create a plausible but fictional story such as the following: "During college I had the opportunity to complete original research with a biochemist who was synthesizing new organometallic compounds. Key biomolecules such as hemoglobin in blood and chlorophyll in green plants are organometallic compounds. (If desired, project images of these two molecules that contain an iron or magnesium atom, respectively, in the center of a complex organic "cage"—see Internet Connections: Wikipedia) Unbeknownst to my biochemist mentor, I actually succeeded in creating a new life form! Through the years, I've kept samples alive in a state of suspended animation, and today, I've brought in a container of specimens to show you." Pull out a taped-over, camouflaged tube of Duco Cement. Ask the learners to generate multiple hypotheses about how your organism might react to light and why.

2. *Explore* (participatory demonstration-experiment)*:* With great fanfare, place one "organism" (drop) on a blank transparency on an overhead projector or under a document camera. As the learners respond unenthusiastically to the inanimate drop, ask them to make and record observations that support or refute the hypotheses of living, dead, or never was living. As most learners will equate the drop's lack of motion as a sign that it is dead or never was living, challenge the assumption that all living organisms move. Hold up and read the cover of a packet of baker's yeast (i.e., "active" dry yeast) and ask if yeast is a form of life. If desired, place some yeast in petri dishes to pass around for examination. (*Note:* Do not pass around the drop of glue, as students will notice the characteristic smell.) Ask students to list critical environmental requirements that might be missing that could be added to help yeast or the mystery organism come out of a possible state of suspended animation. Note that living organisms can only live in ecosystems that provide the raw materials and environmental conditions that support their life processes.

   Add water to a petri dish, place it on an overhead projector (or under a document camera), and again add a fresh drop of the unidentified "mini-monster" specimen. Ask the learners to record their observations. Elicit from the learners a list of distinguishing characteristics of living things. Do all organisms need to assimilate or import food, or can some organisms make their own food from simple chemical compounds? Based on your observations, is this organism more likely to be an autotroph or a heterotroph, and why? After recording their ideas, ask if there are any objects or variations they'd like to add to the creature's environment to test how it would respond. Perform some of these additions (e.g., wood shavings or pepper) as time permits and ask additional questions, such as the following:

- What are some of the ways organisms of the same or different species can react?
- Are males and females of the same species always distinguishable by outward appearances? If desired, one drop can be made to look egglike and another spermlike depending on how the drop is released.
- Is there such a thing as a macroscopic, single-celled organism?
- Why are there limits on cell size?
- If we decide the specimen is living, would this be an observation or an inference? Why?
- How might we account for the decrease in activity of the "drop" over time? Have teams design experiments to support or refute the "living" hypothesis.

Note: Explore-phase activities typically include hands-on explorations by students. If desired and time permits, Extension #1 can be used to involve students directly in microscopic investigations of *Vorticella*. Other organisms such as amoeba and paramecium can also be explored in readily available laboratory activities (see Internet Connections).

3. *Explain:* In science, experimental answers to even such deceptively simple questions as "Is it living or nonliving?" provide the basis for the next round of questions and explorations. The "drop" is most certainly not alive! But its cell-like shape and motility can be used as an entry into discussing how the characteristics of life reveal the unity that underlies the immense diversity of micro- and macroscopic life forms. Supplement standard textbook readings with a PowerPoint slide show, video, and/or internet-based explorations that feature this diversity. Videos and

animations have the advantage of featuring the intra- and inter-cellular movement and 24/7 activities that characterize all life at the microscopic, cellular level, whether or not the macroscopic organism seems to be moving or doing anything.

The three Points to Ponder quotes can be used to introduce the idea of "nested worlds within worlds" and excite students about subsequent lessons on microbiology. Historically, the development of the microscope (as with its earlier cousin the telescope [Panek 1998]) "opened the eyes and blew the minds" of scientists and redefined what counted as scientific observations and evidence. Previously unknown worlds (i.e., undiscovered scales of reality) could be explored without traveling across the seas to distant lands. But analogous to geographic explorations of "uncharted waters and the new world," these new voyages of discovery created a paradigm shift that affected all areas of human culture and our sense of our place in the universe. Be sure that students come to understand this big picture ("forest") perspective, along with subsequent lessons on the names and functions of the various parts of the microscope (or "trees"). If desired, shortened forms of Activity #6 or Activity #9 can serve to transition from observations with the naked, unaided eye to using hand lenses, then 30–100× handheld microscopes, and finally compound microscopes (in Extension #1).

Note: The actual chemical explanation of the FUNomenon of how Duco Cement interacts with water is not immediately relevant to the focus on the biological characteristics of living things. But, if desired, interested high school students can be given a hint that it involves surface tension effects between the nonpolar organic chemicals in the Duco Cement glue and the polar water molecules. They might wish to explore the reactions of other glue brands that are found to work to determine any commonalities in ingredients and the effect of a drop of detergent on the movement of the glue mini-monster or how it behaves in corn oil, alcohol, or other liquids.

Safety and Cleanup

The water can be washed down the drain and the congealed glue can be easily separated and thrown into a trash can.

4. *Elaborate:* Follow-up macro- or microscopic, hands-on explorations can be designed and performed with food-grade yeast to discover that unlike the glue mini-monster, yeast is truly alive

(Internet Connections: Access Excellence, American Society for Microbiology, Exploratorium, GENE Project, Red Star Yeast, and Serendip). Experimental variables for testing include various potential foods (e.g., equal masses of mono-, di-, and polysaccharides; confectioners, brown, and table sugars and artificial sugar substitutes; or equal volumes of various fruit juices and carbonated soft drinks) and/or reaction temperatures either above or below those listed in bread recipes. The carbon dioxide released from a closed container (with a one-hole stopper opening) can be counted as bubbles directed via Tygon tubing into a solution of bromothymol blue (BTB) indicator. BTB will change color from blue to green to yellow with an increasing concentration of carbon dioxide (and decreasing pH values) over a 40- to 60-minute test. Yeast can remain in suspended animation for long periods of time until the proper environmental conditions (water, warm temperature, and food) activate it to "come alive" and it begins to metabolize (producing carbon dioxide and alcohol as waste products). *Note:* High-power (400×) magnification is needed to see individual yeast cells.

Alternatively, student teams can complete library or internet research projects on the work of early microbiologists such as Antonie van Leeuwenhoek (who designed single-lens microscopes with magnification powers up to 300–400×) and Robert Hooke (who made lower magnification but true compound lens microscopes). The Extension activities and Internet Connections serve as entry points to other small life forms that can be studied in the lab (e.g., brine shrimp and paramecium).

5. *Evaluate:* Written or oral reports of the Elaboration phase can be graded or students can develop creative presentations (e.g., posters, demonstrations, analogies, PowerPoint slideshow, etc.) to summarize the characteristics and requirements for life.

   As an alternative summative assessment, demonstrate a toy, battery-powered "smart" car (i.e., it "responds" to its environment by changing its direction when it encounters barriers) as a visual model for real cars that carry smaller, more fragile, "symbiotic life forms" (humans). Ask students to list all of the characteristics of life that human-driven cars display (or lack) that

might lead an alien scientist to hypothesize that the car-human combination is alive. What tests might they perform to disprove this hypothesis? See the Martian and the Car Story (Internet Connections: Biology Corner).

Debriefing

When Working With Teachers

The characteristics of and requirements for life, like many concepts in science textbooks, are often presented in a didactic, "tell 'em, don't show 'em" manner that fails to help students construct meaningful and memorable connections. The result is that though students may be able to regurgitate a list of words in the short term, they fail to gain any lasting, long-term understanding of important concepts and conceptual themes (i.e., they "miss the forest for the trees" and the "wow for the words"). Students come to see the world in black and white, rather than in shades of gray or, better yet, a continuum of colors. Things that fall between the lines of our rigid classification systems (e.g., viruses: living or nonliving?) are presented as anomalies of nature rather than artifacts of our attempts to provide simplified explanations of nature. Similarly, all-or-nothing thinking limits students to "thinking within the box" of our definitions (e.g., Do *all* living things display *all* the characteristics of life at *all* times?).

Teaching science in a noninquiry, "it's true because the textbook, teacher, and tests say so" fashion creates major misconceptions in students' minds about the content and nature of science. Also, analogous to the drop of glue "out of water," students may appear mentally and emotionally "dead" when placed in noninteractive, FUNomena-deprived learning environments. Also, point out the value of using historical quotes and stories to help put a human face on science and give students a sense of the excitement of original breakthroughs and discoveries (see Extension #1 and Internet Connections: ActionBioscience.org). Appendixes A, B, and C can also be discussed in the context of this 5E Teaching Cycle.

When Working With Students

This series of activities is an "engaging" way to introduce a mini-unit on the characteristics and requirements of life (or as a type of extended Engage phase) just before introducing the cell theory unit. It also can be used to introduce the idea of the ecological interactions between an organism and its environment (or *systems*) and the unity (e.g., membrane-bound cells and genetic material) and diversity of life. See also Activities #6, #10, and #12.

Extensions

1. *Carnival of Cavorting, Cellular-Sized Creatures:* Students can examine the variety of vertebrate, invertebrate, and microscopic pond life visible at progressively greater levels of magnification provided by the unaided eye, a 4–10× hand lens, a 30–100× handheld microscope, and a 100–400× compound microscope (e.g., see the American Society for Microbiology, Microbe Zoo and Molecular Expressions websites in Internet Connections). They can relive the excitement of the discovery of "microworlds" by repeating Leeuwenhoek's observations (dated October 9, 1676) of the "cavorting, wretched beasties" we now call *Vorticella*. These unicellular protista can be found attached to duckweed floating in pond water or may be ordered as a single species culture from a science supply company. *Vorticella* can be observed in groups at 100× and individually at 400×. Students can compare their written observations and sketches to Leeuwenhoek's description:

    > When these animalcules bestirred 'emselves, they sometimes stuck out two little horns, which were continually moved, after the fashion of a horse's ears. The part between these little horns was flat, their body else being roundish, save only that it ran somewhat to a point at the hind end; at which pointed end it had a tail, near four times as long as the whole body, and looking as thick, when viewed through my microscope, as a spider's web … These little animals were the most wretched creatures that I have ever seen; for when … they did but hit on any particles or little filaments … they stuck intangled in them; and then pulled their body out into an oval, and did struggle, by strongly stretching themselves, to get their tail loose; whereby their whole body then sprang back towards the pellet of the tail, and their tails than coiled up serpent-wise, after the fashion of a copper or iron wire that, having been wound close about a round stick, and then taken off, kept all its windings. (Dobell 1958, pp. 118–119)

Note: Leeuwenhoek himself never published a scientific paper or a book; all of his observations were described in letters (approximately 200 of them) to the Royal Society in London. Given the strange microworlds he described and his lack of scientific credentials, his observations were initially met with a mixture of microphobia and skepticism. His English contemporary, Robert Hooke, working at lower magnification scales, confirmed the gist of Leeuwenhoek's letters (see Internet Connections for facsimile copies of both of their works).

2. *Brine Is Fine If You're a Shrimp!* Various laboratory investigations can be explored with brine shrimp (see Internet Connections: Brine Shrimp Labs). So-called sea monkeys are a species of brine shrimp developed and marketed as a novelty science exploration kit in the 1960s. Known by the scientific name of *Artemia NYOS,* they grow from the size of a period to 1/2 to 3/4-in. within 4 weeks. As the name suggests, brine shrimp are adapted to live in saltwater environments.

3. *Cinematic Small-Scale Scenes:* Various books and movies such as *Horton Hears a Who* (1954 Dr. Seuss book and 2008 film) and *Honey, I Shrunk the Kids* (1989) provide humorous, fictional entry points into the world of life in the just-out-of-sight range. *The Secret Life at 118 Green Street* (1994; 50 min.) is an engaging science film that uses macrophotography and scanning tunneling microscopy to expose the amazing array of unseen microfauna that go about their daily business of eating and reproducing in our neat and clean homes, countertops, bed linens, walls, and bodies without our noticing. This film is based on the book *The Secret House* by David Bodanis. See also *More Brain-Powered Science*, Activity #18, for a simulation of the spread of a microbial infection, and the book *Human Wildlife: The Life That Lives on Us,* by Dr. Robert Buckman, who claims, "Your body has 100 trillion cells, but only 10 trillion are human. The rest belong to the bacteria, fungi, viruses, and parasites that live on or in us" (Buckman 2003, back cover). Activity #17, Extension #1, in *More Brain-Powered Science* challenges students to critically explore the mathematics of scale related to this quote.

4. *Virtual Tours of Cellular Circuses*: The Microscopy-UK website offers an online Pond Life ID kit and a Virtual Pond Dip. The Cells Alive website contains a variety of activities, which include scale models that contrast a human thumb, pin, and hair to the microscopic world. See other simulation sites in Internet Connections. Also, the Jonathan Swift poem in the Points to Ponder section (and his book *Gulliver's Travels*) can be used to introduce the relative size scales (i.e., decrease in orders of magnitude) across the series: eukaryotic cells to prokaryotic cells to viruses to biomolecules to atoms to subatomic particles.

5. *MacroLooks at the MicroCosmos*: A variety of books, films, and websites (see Internet Connections) are available to help introduce students to the diverse, exotic microworlds hidden from our naked eyes. Consider, for example, a series of books by the microbiologist Lynn Margulis and her science writer son, Dorion Sagan: *The Microcosmos Coloring Book* (1988), *Garden of Microbial Delights: A Practical Guide to the Subvisible World* (1993), and *Microcosmos: Four Billion Years of Microbial Evolution* (1997).

Internet Connections

- 101science.com: Everything you need to know about paramecium (lab, videos, and information): *http://101science.com/paramecium.htm*

- Access Excellence Activities Exchange:

    Cells (many activities): *www.accessexcellence.org/AE/ATG*

    Living or Nonliving? (radish seeds, brown sugar, brine shrimp eggs, yeast, sand, etc.): *www.accessexcellence.org/AE/ATG/data/released/0067-HopkinsKathryn/index.php*

    Yeast fermentation: *www.accessexcellence.org/LC/SS/ferm_activity.php*

    The Four Creatures (Duco Cement mini-monster, another simulation, and two live specimens): *www.accessexcellence.org/AE/ATG/data/released/0114-LenoreKop*

- ActionBioscience.org: *www.actionbioscience.org/education.*

See Classroom Methodology: Using Case Studies to Teach Science (by Clyde Freeman Herreid) and Tapping Into the Pulse of the History of Science With Case Studies (by Douglas Allchin)

- American Society for Microbiology: Classroom Activities (e.g., Taste Test: Can Microbes Tell the Difference [yeast experiment] and Pond Scum: Investigating Microorganisms): *www.asm.org/ index.php/education/classroom-activities.html*

- Antonj van Leeuwenhoek: Father of Microbiology (his original Letters to the Royal Society): *www.vanleeuwenhoek.com*

- Biology Corner: Cells (analogy, labs, songs): *www.biologycorner. com/lesson-plans/cells*

    The Martian and the Car Story: Characteristics of Life: *www. biologycorner.com/worksheets/martian.html*

    Paramecium Lab: *www.biologycorner.com/worksheets/ parameciumlab.htm*

- Brine Shrimp Labs:

    Access Excellence: Brine Shrimp: Fish Farm Open Ended Lab: *www.accessexcellence.org/AE/AEC/AEF/1996/manerchia_ fish.php*

    American Physiological Society: Little Shrimpers: A Brine Shrimp Activity: *www.the-aps.org/education/k12curric/activities/ pdfs/strong.pdf*

    North Carolina State University: Brine Shrimp Project: *www. ncsu.edu/sciencejunction/downloads/brishrim.pdf*

    Sea Monkeys (brine shrimp): *www.sea-monkey.com/html/ aboutsm/whatarethey.html*

- *BSCS 5E Instructional Model: Origins, Effectiveness and Applications*: *www.bscs.org/pdf/5EFull Report.pdf* (65 pages)

- Cells Alive: *www.cellsalive.com* (see especially the How Big interactive animation)

- Cell Biology Animations: *www.johnkyrk.com*

- Cell-ebration: How cells work: *http://people.usd.edu/~bgoodman/ Cell-ebrationframes.htm*

- Cellupedia (Cell simulation game of cellular evolution, evolution timeline, cell science, etc.): *http://library.thinkquest.org/C004535/cell_simulation.html*

- Dennis Kunkel Microscopy, Inc. Science Stock Photography: *www.denniskunkel.com*

- Disney Educational Productions: *Bill Nye the Science Guy: Cells* ($29.99/26 min. DVD): *http://dep.disney.go.com*

- Diversity of Life: *www.diversityoflife.org*

- Exploratorium: Science of Cooking: Yeast-Air Balloons: *www.exploratorium.edu/cooking/bread/activity-yeast.html*

- Flinn Scientific: Glue Monsters: Are They Alive?: *www.flinnsci.com/Documents/demoPDFs/Biology/BF10227.pdf*

- GENE Project Yeast Experiments: *www.phys.ksu.edu/gene/chapters.html*

- Howard Hughes Medical Institute, BioInteractive (variety of free DVDs): *www.hhmi.org/biointeractive*. See Viral Outbreak: The Science of Emerging Disease and Exploring Biodiversity: The Search for New Medicine.

- MedMotion Interactive Health Care Education: Animations and Images: Cell and Molecular Biology: *www.medmotion.com/gallery.shtml*

- Microbe Zoo: Water World: Pond: *http://commtechlab.msu.edu/sites/dlc-me/zoo*

- Microbus: Pond Water Critters including Ciliates (e.g., *Vorticella*): *www.microscope-microscope.org/applications/pond-critters/pond-critters.htm*

- Microscopy-UK: (2D and 3D images): *www.microscopy-uk.org.uk/index.html*

- Molecular Expressions: Digital Video Gallery, Streaming Video, and Downloads: Pond Life: *http://micro.magnet.fsu.edu/moviegallery/pondscum.html*

- Red Star Yeast Science Experiments: *www.lesaffreyeastcorp.com/SoY/experiments.html*

- Robert Hooke's *Micrographia: Some physiological descriptions of minute bodies made by magnifying glasses with observations and inquiries thereupon* (1665 book with original text and images): *http://archive.nlm.nih.gov/proj/ttp/flash/hooke/hooke.html* and *www.gutenberg.org/ebooks/15491*

- Serendip: Hands-on Activities for Teaching Biology to High School or Middle School Learners:

  See Introduction to Biology: Yeast, Cellular Respiration in Yeast, and other resources: *http://serendip.brynmawr.edu/sci_edu/waldron*

- Stuyvesant High School: Observing Living Protists and Paramecium Laboratory Exercise: *http://bio.stuy.edu/labs/LAB4MN.DOC*

- The Biology Project, University of Arizona (interactive online resources for middle school, high school, and college: tutorials, problem sets, and lesson plans): *www.biology.arizona.edu/site.html*

- Wikipedia:

  http://en.wikipedia.org/wiki/Chlorophylls (2D structural formula image)

  http://en.wikipedia.org/wiki/Hemoglobin (2D structural formula image)

  http://en.wikipedia.org/wiki/Robert_Hooke

  http://en.wikipedia.org/wiki/Leeuwenhoek

Answers to Questions in Procedure, steps #1 and #2

1. Living organism may move toward light for warmth or to capture the energy for photosynthesis. They might move away from light because it could be harmful to them. Or they may not respond to it at all because they are unable to sense its presence.

2. At a minimum, "it" appears to have some kind of boundary layer to define itself as separate from its environment. Its lack of motion will cause many learners to assume it is dead or never was living.

Yeast appears to be either nonliving, dead, or perhaps in a state of suspended animation until the environment includes water, a food source (sugar), a certain range of temperature, and oxygen (although it can ferment sugar in the absence of oxygen).

Note: Baker's yeast, *Saccharomyces cerevisiae,* is one of 1,500 yeast species of eukaryotic microorganisms that are part of the fungi kingdom. Naturally occurring yeasts were the agents responsible for some of the first chemical reactions (after fire) controlled by our ancestors some 6,000–10,000 years ago (i.e., to make bread and the fermentation of sugars in grain to produce ethyl alcohol). Interestingly, Antonie van Leeuwenhoek, who observed them in 1680, did not consider them to be alive. Louis Pasteur's research for the French wine industry in 1857 corrected this misconception.

The "drop" will seemingly come alive when dropped into a petri dish of water. It will move quite rapidly at first and may even appear to have a flagellum (depending on how the teacher makes the drop). It also seems to eject some "waste product" into the water as it moves. Living organisms are characterized by their membrane-bound cellular organization, intracellular and intercellular movement, ability to metabolize food for energy and building material, excretion of waste, growth and development, ability to maintain homeostasis while responding to stimuli and maintaining a dynamic separation from their environments, asexual and/or sexual reproduction, and evolution over time as a species.

Organism can be classified as autotrophs or heterotrophs. The former are most likely photosynthetic and contain the green pigment chlorophyll. Since this "organism" is not green, one could test out various "food" possibilities, including wood and graphite pencil shavings, pepper, and sugar. The organism can interact with other organisms by mating with them (if the same species), eating them, ignoring them, symbiotically assisting each other, and so on. Not all species have genders and reproduce sexually. In cases where there are two genders, they may or may not be easy to distinguish by humans. Most macroscopic organisms are multicellular because most cells are microscopic due to surface

area and volume effects that limit the movement of nutrients and waste materials via diffusion, osmosis, and active transport (see Activity #10, Osmosis and "Naked" Eggs). At this point, any judgment about whether or not the specimen is living would be an inference. If we assumed it was alive, the decrease in movement over time could be explained by an inappropriate food source; a negative reaction to heat, light, or oxygen; or other changes in circumstances. Experimental tests could look for signs of metabolic activity, reproduction, microscopic cellular structure, and other qualities.

Activity 8

Water "Stick-to-It-Ness"
A Penny for Your Thoughts

Expected Outcome

Water (in contrast to other clear, household liquids) assumes and maintains a very distinct semispherical shape when placed on a piece of waxed paper. For related reasons, a discrepantly large number of drops of water can be placed on top of a penny without spilling off. Also, a cork can be made to float either in the middle or along the perimeter of a cup of water, and water can be poured down a long string without spilling. Extension activities contrast the pseudoscience hoax of DHMO (dihydrogen monoxide) to the "strange but true" properties of water, explore a water-cornstarch mixture ("Oobleck"), and consider related biological adaptations.

Science Concepts

The often initially discrepant, macroscopic properties of materials can be explained in terms of interactions within and between submicroscopic molecules. Scientists' experimental and theoretical work involves a constant interplay of hands-on and minds-on "manipulation" of materials and ideas that cut across many orders of magnitude in size. At the scale of molecular models, water is depicted as an electrically neutral but polar, V-shaped (or Mickey Mouse head–shaped) molecule with a partial positive charge on the hydrogen "end" and a partial negative charge on the oxygen "end." This charge separation allows for extensive hydrogen bonding that results in water's high degree of adhesion, cohesion, and surface tension; high boiling point and heat of vaporization; and other unusual macroscopic-level physical and chemical properties for its relatively small molecular mass (18 g/mole) and size.

Water is the most uncommon common substance on Earth, where it exists naturally in all three physical states or phases under ambient temperatures and pressures. Its presence in and biogeochemical cycling through the atmosphere, hydrosphere, and lithosphere make it the defining feature of all living ecosystems from the simplest of single-celled organisms to the Earth's biosphere in its entirety. Yet most humans who live in developed nations take this most essential and precious chemical compound for granted. In addition to challenging students to see water from a new perspective, this series of hands-on explorations features the importance of identifying variables and using controls to obtain experimental consistency and reproducibility. It also points to the power of mathematics and molecular theory.

Science Education Concepts

Science concepts, inquiry skills, and habits of mind can be developed via intentionally designed sequences of FUNdaMENTAL 5E-type activities (Engage, Explore, Explain, Elaborate, Evaluate). Students can experience "strange, foreign lands" such as molecular shape and bond theory in the "familiar world" of everyday materials (e.g., water). However, hands-on "messing about with materials" does not necessarily automatically result in minds-on learning. Scaffolded

116

teacher-student, student-student, and student-phenomena interactions are needed to catalyze mental *engagement* and guide students' subsequent *explorations, explanations,* and *elaborations.* The proximate "right" answers on a particular laboratory investigation (e.g., the number of drops of water that will fit on a penny) are typically of far less relevance than the underlying explanatory theory that is uncovered in exploring a given phenomenon. *Evaluations* of lessons learned need to focus on the latter.

The cohesive properties of water can also be used as a *visual participatory analogy* for effective classroom "ecosystems" where synergistic, bondlike interactions between individuals create emergent properties (i.e., the whole is greater than the sum of the parts). It can also be used to represent the power of teachers' "stick-togetherness" in professional learning communities and interactive networks to further develop their skills in designing cooperative, experientially rich learning environments for their students.

Materials

- A globe, poster, and/or projected internet image or video clip of Spaceship Earth
- Each learner-dyad will need 1 or 2 pennies, 1 or 2 plastic eye-droppers (if desired, these may be cut to different-size diameter openings for different-size drops), a toothpick, several milliliters of water, and one or more other clear household liquids (e.g., corn oil, olive oil, mineral oil, rubbing alcohol [may be either ethanol or isopropanol that varies in concentration from 50–91%], witch hazel, etc.), several granules of black pepper or chalk dust, 2 clear plastic cups, 10 ml and 100 ml graduated cylinders, and a small piece of waxed paper.

Before using, clean the pennies by soaking them for 10–20 minutes in vinegar with a small amount of table salt (do not use soap). After the soak and a light scrub with an old toothbrush, dirt and oils are removed to reveal shiny, "new" pennies. An overnight "bath" in a soda works as well (due to phosphoric and carbonic acids). Hand lenses are optional. A related teacher-demonstration uses dish detergent, small clear plastic cups, and, a 4–10 ft. piece of cotton string.

Points to Ponder

Now there is one outstanding important fact regarding Spaceship Earth, and that is that no instruction book came with it … We are not going to be able to operate our Spaceship Earth successfully nor for much longer unless we see it as a whole spaceship and our fate as common. It has to be everybody or nobody.

—R. Buckminster Fuller (1895–1983), American architect, engineer, philosopher, futurist, and author of the *Operating Manual for Spaceship Earth* (1969)

There are no passengers on Spaceship Earth. We are all crew.

—Marshall McLuhan, Canadian philosopher, futurist, and communications theorist (1911–1980)

On consideration and by the advice of learned men, I thought it improper to unfold the secrets of the arts (alchemy) to the vulgar, as few persons are capable of using its mysteries to advantage and without detriment.

—Tycho Brahe, Danish astronomer (1546–1601)

Procedure

When Working With Teachers

To conserve time but maintain involvement, teams can specialize in different phases and report their results in the order of the 5E Teaching Cycle to the whole group.

1. *Engage: Spaceship Earth: The Water Planet:* Quickly introduce this mini unit by pointing to a globe, poster, or projected internet image of Earth as seen from space and asking the learners to identify Earth's most dominant and unique feature relative to

Points to Ponder *(continued)*

Caroline (one of two female students): To confess the truth, Mrs. B., I am not disposed to form a very favorable idea of chemistry, nor do I expect to derive much entertainment from it. I prefer the sciences which exhibit nature on a grand scale, to those that are confined to the minutiae of petty details … I grant, however, there may be some entertaining experiments in chemistry, and I should not dislike to try some of them … Mrs. B (her teacher): I rather imagine, my dear Caroline, that your want of taste for chemistry proceeds from the very limited idea you entertain of its object … chemistry is by no means confined to works of art. Nature also has her laboratory, which is the universe, and there she is incessantly employed in chemical operations. You are surprised, Caroline, but I assure you that the most interesting phenomena of nature are almost all of produced by chemical powers … A woman may obtain such knowledge of the science as will not only throw an interest on the common occurrences of life, but will enlarge the sphere of her ideas, and render the contemplation of nature a source of delightful instruction.

—Excerpt from *Conversations on Chemistry*, a textbook by Jane Marcet (1769–1858)

our moon and other planets in the solar system. Alternatively, the first few minutes of NASA's *Earth as a System* video clip may be used to make the same point (see Internet Connections: Teachers' Domain). Share the quotes from Buckminster Fuller and Marshall McLuhan, and suggest that to help us become better "crew members," this mini unit will investigate the strange but true properties of water, the most common, uncommon substance.

Shape Shifter and "Stick-Togetherness": Like other fluids, water assumes the shape of the container in which it is placed. Why? Turning this everyday, matter-of-fact observation into a topic for exploration requires looking at water under conditions where it is less restrained.

Ask the learners to predict-observe-explain (POE) what shape they think water will assume when drops (from an eyedropper) are placed on a flat, nonabsorbent surface such as waxed paper, and why. What happens when two drops are placed close together or pulled near to one another with the end of an eyedropper? If the waxed paper is tilted, does a drop of water slide or roll? *Note:* Have the learners place a speck of black pepper or chalk dust on the drop to investigate this last question. Compare water's tendency to "stick together" to that of other clear, household liquids (preferably those with little or no water added). Be sure that learners look at several different types of liquid drops on waxed paper from a sideways, eye-level perspective. Hand lenses can be used to magnify the view.

Challenge the learners to develop their own questions about their observations and/or consider questions such as the following:

- What forces might cause these FUNomenon to occur?
- What might these macroscopic behaviors suggest about the "molecular personality" of water versus other fluids?
- What else do we know about the physical and chemical properties of water?

Let the learners play with these questions and materials to see what they can discover before proceeding to the next exploration. The questions and hypotheses the learners generate are as important as the ones listed above and should be shared with the entire class to prime their thinking. Neither teacher-designed nor student-generated questions should be answered by the teacher during this phase. The key is to let the phenomenon raise questions that students find interesting and that provide opportunities for students to talk about what they already know about the content to provide a window into their thinking for the teacher. As an aside, learners can also explore (a) the effect of a toothpick dipped in detergent on drops of water (or this can be used as an unrevealed "magic" trick on an overhead projector or under a document camera where only the teacher's toothpick has this effect), and (b) the magnifying properties of a water drop and how the shape of the water drop compares to that of their magnifying lenses.

2. *Explore: Water Drops on a Penny: Mathematics and Measurements Matter*

   a. Predict what shape water and the other liquids will assume when drops are placed onto a penny and how many drops of different liquids will fit on a penny without spilling off.

   b. After the learners make and record their initial "wild guesses," challenge them to consider how they could make a more scientifically informed estimate before they empirically test their initial guess. What measurements and formula might be useful? If necessary, prime their thinking with questions such as the following:

      • Would the length of the perimeter (i.e., circumference) or the surface area of the coin have more predictive power?
      • What is the approximate surface area of the penny (and would larger coins such as a quarter allow for more drops)?
      • How high do you think the drops of liquid will stack?
      • What would be the volume of this cylinder of liquid?
      • What is the approximate volume of a drop of liquid?
      • How many drops would fit into this hypothetical cylinder on top of a penny?
      • Will the column be flat at the top (like a can) or curved?
      • If the top is curved, in what direction (concave or convex) does this change with the addition of more drops, and why?

   c. Perform (and observe) this experiment with a partner. Consider the benefits of averaging multiple trials or pooling the class data.

   d. After allowing the teams to make their initial test runs and comparisons to both their own predictions and the actual counts from other teams, challenge them to consider if there are any relevant variables that need to be "controlled for a fair test." If time permits, try other coins with different surface areas and surface features.

   e. What could you do to increase versus decrease the number of drops that would "hold on?"
      Do the other household liquids work as well as water? If not, why? Do the liquids differ in macroscopic ways that might

account for the results? If not, would a molecular-level perspective help?

f. Looking at the penny-liquid system that is near the maximum number of drops from the side, would you describe the shape of the curvature of water on the top of the penny as concave (U-shaped) or convex (an upside-down U)? How could you create a situation where water's surface assumed the opposite shape?

3. *Explain: Concave or Convex: It's Perplexing!* Let the learners play with two different-size graduated cylinders, translucent plastic cups, and water. Remind them to take a sideways, eye-level view of the cup and water to address the following questions:

a. When water is poured into a graduated cylinder, what shape does it take at the surface? Is this curvature more noticeable in a 10 ml or 100 ml graduated cylinder? Why? What forces could be causing the seeming antigravity effect along the inside perimeter of the cylinders?

b. Would these same forces be at work when water is poured into (but not overflowing from) a plastic cup? Is the curvature of water in a cup as noticeable as with the graduated cylinder? Why or why not?

c. If you continue to add water to an already "filled" cup, one drop at a time, what shape will the curvature assume? Why? Alternatively, learners can be challenged to guess (and test) how many paper clips (or pennies) can be added to a "filled" cup without it spilling over.

It is essential to provide opportunities for learners to explore empirical questions *before* prematurely offering theoretical answers. Between the end of the Explore phase and the beginning of the Explain phase, laboratory results should be pooled and critically discussed. Then, building on this foundation of shared experiences, the teacher can formally introduce grade level–appropriate concepts (and assign readings) related to the chemistry and unique properties of water. Over the course of several days of lessons, 3D molecular models, videos, computer animations, and people-as-molecules simulations (Battino 1979) can help students make the connection between molecular-level theoretical explanations (e.g., polar covalent bonds, molecular shape, polarity, and hydrogen bonding) and macroscopic observations (e.g., cohesion, adhesion, surface tension, and the formation of a concave or convex meniscus).

Understanding that "very small, invisible things" (e.g., molecules) are really "big ideas" requires the learner to "manipulate" both materials and ideas. Hands-on explorations are necessary but rarely sufficient (by themselves) to catalyze minds-on learning, especially in cases involving unseen and unfelt entities such as molecules and intermolecular forces. Making such entities "sensible" takes time and the guided questioning and mentoring of a skilled teacher. Expecting students to "leap tall buildings in a single bound" is not realistic since development of the kinetic molecular theory took hundreds of years. Guided reading activities of select passages from a textbook, the internet, or other sources can be supported by teaching students to use graphic organizers such as concept maps. *Note:* Activity #20 in *Brain-Powered Science* and activities #3, #8, and #15 in *More Brain-Powered Science* provide a variety of easy-to-do demonstration-experiments that provide compelling arguments for the idea of Daltonian atoms and molecules that would precede this 5E mini unit and the idea of polar covalent bonding.

4. *Elaborate: Science Is Magical:* Challenge learner-teams to use their experiences from previous mini-experiments to develop and explain a "science magic trick" where a cork can be made to float either in the center or at the perimeter of a cup at their command. What is the underlying principle involved in this trick? Also see any of the Extension activities.

5. *Evaluate: Stringing Water Along:* A dramatic demonstration of water's "stick-to-it-ness" is to pour water from an elevated beaker to a ground-level beaker 4–10 ft. away, along a wet, taut cotton string. The water can be colored with food coloring for greater visibility or a flashlight can be shone on the water trickling down the rope in a darkened room.

Ask:

a. How can we use what we know about water's unique properties to account for this unusual demonstration?

b. Would nylon string work for this demonstration? Could other household fluids be transported this way? Why or why not?

c. What will happen if dish detergent is added to the water? (*Note:* Be sure to keep these detergent-contaminated setups separate from the pure water setups.)

A large number of demonstrations and chemical magic tricks can be explained with the concepts of surface tension and surfactants. For example, pepper or chalk dust will rest on the surface of water until it is touched with an object that has been dipped in detergent; this is easy to demonstrate in a petri dish placed on an overhead projector or under a document camera.

Debriefing

When Working With Teachers

Discuss the idea of using short, minds-on, inquiry-oriented, hands-on explorations on a regular basis to encourage students to "HOE the garden of their minds to support the seeds of new ideas." The historical quotes by Tycho Brahe and Jane Marcet point to a shift in views from chemistry as a science for only the most learned of "men" to chemistry as a "central science" for all. If time permits, the "big idea" of common themes (systems, models, constancy and change, and scale; see Appendix C) in teaching science is worth discussing. Also, water's cohesiveness can be used as a *visual participatory analogy* to discuss the power of a classroom based on cooperation and mutual benefit and to acknowledge teachers' "stick-to-it-ness" as professionals in participating in ongoing lifelong learning and collegial networking.

When Working With Students

Emphasize that principles of science are at work all the time in our everyday experiences, not just in school science labs. For high school students, the underlying science and real-world applications of water's polarity and hydrogen bonding would be further developed via additional activities. Interdisciplinary applications include biology (e.g., insects that can walk on water, capillary action in xylem in plants, etc.), chemistry (e.g., water as a "universal" solvent and the surface-tension-breaking power of detergents), Earth science (e.g., rain drops and the movement of water through different kinds of soils), and physics. Magnetic molecular kits (e.g., H_2O, $NaCl$, etc.) that model electrostatic attractions within and between molecules and ionic solids are available from 3-D Molecular Designs (414-774-656; see Internet Connections).

Use the quote from Tycho Brahe and the *Conversations on Chemistry* book excerpt to activate and challenge misconceptions about the scope of chemistry and its role in our world, as well as the mistaken notion that chemistry is a science "for men only." The book's author and all three of the book's fictitious characters were females studying chemistry in the early to mid 1800s. This book is credited with exciting Michael Faraday's interest in chemistry when he chanced upon it as an apprentice bookbinder (before he was discovered by Sir Humphrey Davy).

Extensions

1. *DiHydrogen MonOxide (DHMO) Hoax:* An engaging STS activity prompt that features the problems of pseudoscience and chemophobia was created at the University of California, Santa Cruz, and has been distributed and extended via faxes and e-mails since the 1990s. Water under an obscure acronym-type name is presented as a health and environmental risk. Photocopied for students (or teachers) to read, this hoax is a great discussion starter about the "uncommon nature of this common substance" and the importance of science literacy. See the Internet Connections for variations on the "DHMO science news release" and the contrasting scientifically valid information on the actual physics and chemistry of H_2O molecules. If desired, view the variety of YouTube videos with actor-scientists recounting the dangers of dihydrogen monoxide. These intentionally fraudulent "news" releases can be contrasted to real studies by the United Nations that report that more people die from polluted water (approximately 3.7% of all deaths) than from all forms of violence, including war (see Internet Connections). Additionally, it has been estimated that more than half of the world's hospital beds are filled with people suffering from water-related illnesses. Water conservation and sanitation efforts are essential to quality of all life on Earth, the water planet.

2. *Variable and Vexing Volatility: It Takes Alkanes (for high school chemistry classes):* Volatility of a liquid is inversely proportional to both molecular mass (i.e., lower-mass molecules evaporate more quickly than heavier ones at the same temperature and pressure) *and* intermolecular bonding (i.e., asymmetrical, polar covalent molecules with higher intermolecular attraction evaporate more slowly than nonpolar ones). Consider the group VI oxides:

- hydrogen oxide/water (mol wt = 18.0 and BP = 100°C)
- hydrogen sulfide/H_2S (mol wt = 34.08 and BP = -60.7°C)
- hydrogen selenide/H_2Se (mol wt = 80.98 and BP = -41.3°C)
- hydrogen telluride/H_2Te (mol wt = 129.63 and BP = -2°C)

With the exception of water, boiling points increase with increasing molecular mass. Water's extensive hydrogen bonding allows it to uncharacteristically exist in all three phases (but predominantly as a liquid) at the typical range of ambient Earth temperatures and pressures.

Though the above series of molecular compounds cannot be readily examined in a classroom, a similar series, the nonpolar hydrocarbon alkane series (methane [CH_4], ethane [C_2H_6], propane [C_3H_8], butane [C_4H_{10}], pentane [C_5H_{12}]...), can be. The progression is from methane, which is a gas under typical room temperature and pressure; to butane, which becomes a liquid when placed under slight pressure (e.g., as in a butane lighter); to pentane, which is a liquid; to increasingly less volatile liquids and eventually to solids. Similarly, within a series of related alcohols such as methanol (CH_3OH: 32 g/mol and BP = 64.96°C), ethanol (CH_3CH_2OH: 46 g/mol and BP = 78.5°C), propanol (exists as two 60 g/mol isomers: 1-propanol, $CH_3CH_2CH_2OH$, with BP = 97.4°C; and 2-propanol or isopropanol, $CH_3CH(OH)CH_3$ with BP = 82.4°C), the volatility decreases (and boiling point increases) with increasing molecular mass. The alcohol series can be investigated in *a teacher demonstration only* by comparing how quickly the liquid left behind from lightly dampened cotton balls evaporates from a chalkboard or other surface. Students should be challenged to predict-observe-explain the sequence.

When comparing a series of dissimilar molecules, the strengths of intermolecular attractions become a significant factor in determining volatility. Consider, for instance, methane/CH_4 (16 g/mol and BP = -161.4°C), water/H_2O (18 g/mol with a BP = 100°C), ethanol/CH_3CH_2OH (46 g/mol and BP = 78.5°C), and acetone/CH_3COCH_3 (58 g/mol with BP = 56.5°C). It is easy to demonstrate on a chalkboard that acetone (with the highest molecular mass) evaporates much faster than ethanol, which in turn evaporates before the water (which has the lowest molecular mass of the three liquids). Acetone is symmetrical and nonpolar, ethanol has some hydrogen bonding, and the V-shaped, polar covalent water molecule has extensive hydrogen bonding. Water's unique chemical and physical properties are all related to hydrogen bonding.

3. *Baffling Biological Behaviors:* Various organisms that are denser than water can "walk on water" due to the hydrophobic nature of their legs and feet, the way their weight is distributed, and the surface tension (or membrane) at the surface of water. See the Internet Connections as starting points for exploring this unusual biological adaptation. Through millions of years of evolutionary "research and development," living organisms have discovered or invented a number of unique ways to apply principles of chemistry and physics to provide selective advantages that help them produce or find and capture food (or escape from being food) and reproduce and support their offspring (see Activity #9). Water makes up the majority of the weight of most organisms (and every living cell). Whether or not an organism lives in water, its relationship to water is a defining characteristic of its life.

4. *Oobleck and Other Non-Newtonian Fluids:* Water and most fluids have a characteristic constant viscosity at a given temperature and pressure. However, non-Newtonian fluids (e.g., cornstarch in water, ketchup, toothpaste, paint, etc.) have viscosities that vary with applied stress and can shear and act like solids over short periods of time. A mixture of cornstarch and green-dyed water can be linked with the theme of environmental hubris, as represented in the classic Dr. Seuss story *Oobleck*. Students can make and investigate the unusual properties of this mixture and view discrepant video clips of "mere mortals walking on water" (e.g., featured on episodes of *Numb3rs* on CBS, *Mythbusters* on

Discovery Channel, and talented "amateurs" on YouTube). Students can also make a homemade version of silly putty with Elmer's glue, borax, and water. See below for directions.

Internet Connections

- 3-D Molecular Designs: *www.3dmoleculardesigns.com/news2. php#water*

- Biological adaptation to water's high surface tension (video clips):

   Basilisk lizard: *http://en.wikipedia.org/wiki/Basilisk_lizard*

   www.societyofrobots.com/robot_jesus_lizard.shtml

   Water Strider:

   http://en.wikipedia.org/wiki/Water_strider

   www.fcps.edu/islandcreekes/ecology/common_water_strider.htm

- *BSCS 5E Instructional Model: Origins, Effectiveness and Applications:*

   www.bscs.org/pdf/5EFull Report.pdf (65 pages)

   http://bscs.org/pdf/bscs5eexecsummary.pdf (19 pages)

- Cell Biology Animations: Water: *www.johnkyrk.com/H2O.html*

- DiHydrogen MonOxide (DHMO) Hoax: *www.dhmo.org*

   http://urbanlegends.about.com/library/bl_ban_dhmo.htm

   http://en.wikipedia.org/wiki/Dihydrogen_monoxide

   ww.snopes.com/science/dhmo.asp

   www.stevespanglerscience.com/experiment/dihydrogen-monoxide

- Disney Educational Productions: *Bill Nye the Science Guy: Fluids* ($29.99/26 min. DVD): *http://dep.disney.go.com*

- Drops on a Penny activity: online lab write-ups:

   Katz, D. A. 2004. *Drops of Liquid on a Penny*: *www.chymist. com/drops%20on%20a%20coin.pdf*

   Trimpe, T. 1999. How many drops of H_2O can fit on a penny? *www.sciencespot.net/Media/pennylab.pdf*

Biology Corner: Scientific Method (Penny Lab, Water in Living Things, and other labs): *www.biologycorner.com/lesson-plans/scientific-method*

- Earth, the Water Planet images: Search Google for photos such as those found at:

    www.epa.gov/glnpo/active/2004/apr03.jpg (full Earth)

    www.urbansprout.co.za/files/images/spaceship_earth.jpg (Earth from the Moon)

- HyperPhysics, Department of Physics and Astronomy, Georgia State University:

    Surface tension: Molecular explanation and applications: *http://hyperphysics.phy-astr.gsu.edu/hbase/surten.html*

- National Geographic Society: Special Issue, April 2010: Water: Our Thirsty World: *http://ngm.nationalgeographic.com/2010/04/table-of-contents*

- Non-Newtonian Fluids:

    http://en.wikipedia.org/wiki/Non-Newtonian_fluid

    www.wisegeek.com/what-is-a-non-newtonian-fluid.htm

    www.answers.com/topic/non-newtonian-fluid

    Oobleck: The Dr. Seuss Science Experiment: *www.instructables.com/id/Oobleck*

    Steve Spangler Science: Cornstarch Science and Quicksand Goo (includes video of "walks"): *www.stevespanglerscience.com/experiment/00000088*

    Silly Putty (homemade):

    www.science-house.org/CO2/activities/polymer/sillyputty.html

    http://chemistry.lsu.edu/webpub/demo-2-silly-putty.pdf

    www.stevespanglerscience.com/experiment/00000039

    Walking on Water (cornstarch mixture): *www.youtube.com/watch?v=f2XQ97XHjVw*

- Official M.C. Escher Website (*www.mcescher.com*) sells a work titled *Waterfall Cascade* (showing an impossible arrangement), which makes for a great prop for an extended discussion of

the unusual and seemingly discrepant physical and chemical properties of water.

- San Diego State University: Biology Lessons for Prospective and Practicing Teachers:

    Properties of Water: *www.biologylessons.sdsu.edu/ta/classes/ lab1/index.html*

- Teachers' Domain: Earth as a System (5:31 min. NASA video): *www.teachersdomain.org/resource/ess05.sci.ess.earthsys.hologlobe*

- United Nations: *Sick Water: The Central Role of Wastewater Management in Sustainable Development*: *www.unwater.org/ downloads/SickWater_unep_unh.pdf*

- U.S. Geological Survey: Water Science for Schools: *http:// ga.water.usgs.gov/edu/mwater.html*

- Wikipedia: *http://en.wikipedia.org/wiki*. Search topics: detergents, surface tension, surfactants, and water and water properties.

Answers to Questions in Procedure, steps #1–#5

1. *Engage:* The Earth globe, image, or video clip all feature the presence of lots of water in all three phases: ice/lithosphere (or technically the cryosphere), liquid/hydrosphere, and gas/atmosphere. When playing with water, the polarity of water molecules will cause water to assume a semispherical shape on waxed paper as it is attracted more to itself (cohesion) than to the nonpolar wax. Note that in a microgravity environment, water would assume a spherical shape, as this shape has the smallest surface area for a given volume. When two drops of water are brought close together, they will be attracted to each other to form a single drop with less total surface area than the two individual drops. If a speck of black pepper or chalk dust is placed on the drop, the drop can be seen to roll (rather than slide and glide) on tilted waxed paper. Other household liquids spread out more when placed on waxed paper because their internal cohesive forces are not as strong as in water. Hydrogen bonding between adjacent water molecules causes water to have a "split personality" as a

polar covalent molecule. This allows it to dissolve an array of substances, from salts (ionic) to common covalently bonded substances such as sugar. Students are not expected to know all of this at the start of the 5E.

2. *Explore:*

   a. Most learners will significantly underestimate the number of drops of water that will stay on top of the penny because they are unlikely to consider its cohesive property.

   b. The approximate surface area of a penny can be calculated from its diameter (approximately 2 cm) and the formula $SA_{circle} = \pi r^2$. Therefore, the $SA_{penny} = \sim 3.14 \text{ cm}^2$. Note that this does not take into account the fact that the penny is contoured rather than flat, and newer pennies may have more prominent, less worn perimeter edges. A reasonable guess for the height of a liquid stack might be 0.2 cm. Learners could then count how many drops of liquid are in 1 cm^3 or 1 ml (i.e., for water and a typical eyedropper, this is about 20 drops/cm^3) and calculate the $V_{cylinder} = 3.14 \, r^2 h \, (= \sim 0.628 \text{ ml})$ to come up with an estimate of about 13 drops. This assumes the cylinder would be flat on the top; in fact, it will be curved, so perhaps 13 drops may be a low estimate.

   c. Actual sample results will vary with the size of the drops, but the relative ranking is water (30–50 drops), witch hazel (about the same as water), 95% pure ethanol (20–30 drops due to less extensive H-bonding), olive oil and corn oil (20–25 drops), mineral spirits (2–3 drops due to limited or no H-bonding or molecular polarity), with corresponding smaller half-sphere buildups. If different concentrations of rubbing alcohol are used (e.g., 91% versus 70% ethanol or isopropanol), those with a higher water concentration will behave more like water.

   d. Variables that could affect how many drops of liquid a penny would hold include the penny's age, wear and tear, how clean the penny is, the relative size of the drops, the height from which the water is dropped, which side of the penny is used, whether the penny is from the United States or Canada, and so on. Of course, the nature of the specific liquid is a very important variable.

e. The more defined the edge on the penny and the smaller the size of the drops, the more drops will "hang on." A drop of detergent will greatly reduce surface tension and far fewer drops will be able to be placed on the penny. (Keep these "contaminated" pennies and droppers separate from the others.) The other liquids do not "stack up and hold together" as well as water. To explain the different results with the different clear, colorless liquids, one needs to take a molecular perspective. Water has the highest cohesive and adhesive forces of any common liquid due to its extensive hydrogen bonding between polar covalent water molecules. Thus, even the mathematics-based estimates (that do not take into account molecular theory) will be incorrect.

f. The water on top of the penny assumes a convex in contrast to the concave shape it takes when it is put into most containers. For instance, we read the volume of water in a graduated cylinder from the lowest point of its concave shape (or meniscus). Unless a cup of water is carefully "overfilled," water will take on a concave shape there as well, although it is only in narrow diameter containers that this is easy to directly observe.

3. *Explain:* Unless a container is carefully overfilled, water will assume a concave or U-shape at the surface that will be more noticeable the smaller the diameter of the cup (or cylinder). Textbooks and online resources explain that the meniscus is due to the combination of intermolecular cohesive forces in water being greatest at the center and water-container adhesive forces being greatest at the perimeter of the cup. The latter effect is most noticeable in thin tubes, where it is called capillary action. However, as the previous drops-on-a-penny activity demonstrates, water's cohesive properties enable a cup to be "overfilled," in which case the curvature shifts to a convex, upside-down U-shape. A cup that appears filled with water will assume a slightly convex shape if more water is carefully added drop by drop or if small objects are dropped into the water. For instance, a surprisingly large number of paper clips (or pennies, gently released edgewise into the surface of the water) will fit into such a cup as the water's cohesiveness enables it to bulge upward into a convex shape that rises noticeably above the lip of the cup.

4. *Elaborate:* The cork "falls downhill" to a region of lowest gravitational potential energy at the middle of a partially filled cup of water with a standard, concave meniscus. But the cork falls downhill to the perimeter (or lowest point) of an overfilled cup with a convex surface.

5. *Evaluate:*

   a. The combination of cohesive forces (water molecules sticking to each other) and adhesive forces (that cause water to cling to the cotton fibers) enables the pouring to proceed without spilling any water if it is not poured too fast and the string is kept taut.

   b. Neither a nylon string nor gasoline will work. In the first case, water molecules are not as attracted to nylon as to cotton, and in the second case, gasoline is nonpolar and will not be attracted to the cotton.

   c. Detergents break the surface tension of water by decreasing its "stick-togetherness." Water containing a drop of detergent will not be able to stack up as high on a penny, nor will this mixture be able to overfill a cup (by adding drops of water or paper clips or pennies to the cup).

Activity 9

Burdock and Velcro
Mother Nature Knows Best

Is burdock a PROBLEM or a SOLUTION?

Expected Outcome

Velcro is explored as an example of a human-engineered invention that was a "copy-cat" inspired by a naturally evolved, "bio-engineered" seed distribution innovation.

Science Concepts

Scientists and engineers use microscopes and other technologies to study, disassemble, and *model* complex *systems* to discover how they work (i.e., how the parts interact to create new, emergent properties) and how to fix broken systems and/or create new, improved ones. Genetic variation, mutations, natural selection, and evolution have "encouraged" living organisms to invent and continually adapt creative solutions to a host of survival problems over 3.8 billion years of research and development (R & D), product field testing, and re-engineering. Even the best natural solutions may eventually fail to meet the challenges of ever-changing environments and become "fossilized failures" (due to a mismatch between *form* [structure] and *function* [purpose] given changed constraints and conditions in particular environments). But extant living organisms represent an immeasurable wealth of lessons learned and are therefore rich quarries to be mined for invention ideas.

Velcro (or generic versions of this hook-and-loop fastener system) is an example where nature "got there first" and an observant human engineer serendipitously discovered and adapted a pre-existing system solution (i.e., burdock–animal fur for seed dispersal) to a seemingly different problem by making use of a new, human-engineered material (i.e., nylon, a synthetic substitute for natural silk). The Elaboration phase of this 5E Teaching Cycle introduces other examples of analogous nature-human inventions in the field of study known as bionics, biomimicry, or biomimetics. Thinking in terms of other living species doing R & D work and drawing analogies between natural and human-engineered "product designs" will be discrepant for most students and many teachers.

Science Education Concepts

This hands-on exploration (HOE) is an example of a mini 5E Teaching Cycle (BSCS's Engage, Explore, Explain, Elaborate, Evaluate) that can be shared with teachers in a relatively short time period to introduce the idea of a 5E cycle as a way to organize multi-day and unit-level planning. 5E Teaching Cycles are likely to be "discrepant" for some teachers in that this curriculum-instruction-assessment (CIA) model emphasizes *engaging* learners' prior knowledge and motivation and *exploring* interesting questions before prematurely offering teacher or textbook, terminology-heavy *explanations* or answers. The Velcro story of engineered alignment of form to function is both a direct example of and a *visual participatory analogy* for the idea of an intelligently designed, integrated CIA system. "Smart" CIA creates synergy between the separate components and a final product (or whole) that is greater than the sum of the separate parts (see Appendix B).

Materials

- Velcro or generic hook-and-loop fastening tape can be cut into approximately 5 cm long sections for each group of 2 or 3 learners that also have either 5–10× hand lenses or 30–60× handheld microscopes and samples of 1 or more types of natural burdock (biennial thistles in the Genus: *Arctium*), cocklebur (Genus: *Xanthium*), sticktight (Genus: *Bidens*), or other natural burr-type seeds that are dispersed by attaching to animals (collect these from a local field-type environment and store for reuse if desired).
- *Optional:* projection video camera and videos or DVDs (see Internet Connections)

Points to Ponder

In the works of Nature, purpose, not accident, is the main thing … Nature does nothing uselessly.

—Aristotle, Greek philosopher (384–322 BC)

Necessity is the mistress and guide of nature. Necessity is the mistress and artificer of nature, the bridle and eternal law.

—Leonardo da Vinci, Italian polymath (1452–1519)

Chance favors the prepared mind.

—Louis Pasteur, French chemist and microbiologist (1822–1895)

Necessity is the mother of invention.

—Anonymous

Procedure

Teachers can be taken through a cut-time version of the first three phases of the following 5E sequence. Share the big-picture focus questions for this mini unit: How do the structures (or *forms*) of human-designed products and those constructed by other living organisms relate to their *functions*? How can all species be considered engineers? These same questions can be shared at the start of the mini unit with grades 5–12 students or reserved until the Explain phase. *Note:* When used with grades 5–12 students, this entire sequence could take as long as a week and could serve as an Engage phase for a larger 5E-based unit dealing with plant reproductive adaptations as part of a yearlong theme of biological adaptations and evolution.

1. *Engage: Clothing Conundrum: Just a Human Problem?*

   Human clothing needs to be secure from unintended, accidental separation during movement, but also capable of being undone when it needs to be loosened or removed. What are some human-engineered solutions to this need for a semipermanent way of attaching and removing different types of clothing? Do a quick whole-class brainstorm by having the learners examine their own (outer) clothing for examples. If no one mentions Velcro, pull apart two attached Velcro strips out of sight and see if the distinctive sound is a sufficient memory cue. In any case, be sure to visibly demonstrate the operation of Velcro using an actual clothing item (e.g., sandals, tennis shoes, pants, jacket, etc.). Briefly discuss the following:

   a. How items such as snaps, zippers, and Velcro are examples of or *models* for *systems* "where the whole is greater than the sum of the parts." That is, interactions between differently designed parts synergistically create something with new emergent properties and different functional capabilities than the individual, isolated parts.

   b. How the same universal scientific laws and principles apply across all levels of reality of both natural and human-designed systems.

2. *Explore: Necessity: The Mother of Invention*

   a. *Clothing Close-up: Views of Velcro:* Challenge individual learners to do the following:

      (1) Predict and draw a sketch of what the two halves of Velcro would look like if examined up close, say from the perspective of a pet mouse. Consider the connections between form (or structure) and function (or purpose) and Velcro's relative advantages or disadvantages compared to other types of fasteners. Have dyad or triad teams share their ideas and develop a best-guess composite model that could account for Velcro's hold-and-release properties.

      (2) Have the teams observe and draw a second sketch of the two Velcro halves using the hand lens or 30× handheld microscope to see if they can use the structures (or forms) revealed at this slightly magnified scale to explain how Velcro functions. Discuss how the two sets of drawings compare and the value of microscopes to extend human sensory limitations.

   b. *Plant Parenting Problem:* Ask the learners to role-play being a plant and discuss the following survival questions:

      (1) Like many plants, I reproduce sexually and produce seeds that contain embryonic plants. What problem would I create for myself (and my offspring) if I simply release and drop my seeds directly below my own location?

      (2) Like many plants I am "rooted in one place" and cannot move to another location; therefore, I need a means of transporting my offspring seeds to colonize distant sites. What different solutions allow plants to transport their seeds ("to get the kids to leave home")?

      (3) Which of these solutions appears to be somewhat analogous to Velcro?

         After briefly discussing the above questions, shift to a hands-on exploration to help discover how burdock-type seeds' surrounding structures (or forms) enable them to serve the function of helping the seeds "break out on their own." Have the learners use a hand lens or 30×

microscope to examine and make magnified-view drawings of sticktight, cocklebur, or burdock seeds and compare them to their previous sketch of the two halves of Velcro.

(4) How are these natural seed carriers similar to or unlike Velcro? What might the "missing half" of this particular type of natural seed dispersal system be? Ask the learners to use the handheld microscopes to examine their own hair (and/or cat fur or wool) and various types of clothing fabric to answer these questions empirically.

3. *Explain: Science Stories: Making Connections*

Further discuss the analogy between the human-engineered Velcro and the natural, bio-engineered system that uses an attach, transport, and release mechanism to disperse seeds. The connection between the forms and functions of these two systems can be discussed in light of the following historical vignette:

In 1948, Swiss engineer George de Mestral returned from a hike in the Alpine woods to find cockleburs clinging to his dog's fur and his wool trousers. Intrigued by this common annoyance of nature hikers, he decided to investigate the phenomenon under the microscope. He found that the cockleburs were covered with a maze of burrs or hooks that would become ensnared in the loops of clothing fabrics or animal fur. The plant had an efficient means of transporting its seeds by using unknowing animal carriers!

De Mestral immediately saw it as a model for a potential new type of fastener. After eight years of development, on September 13, 1955, he patented "Velcro" (French/English: *vel*ours/velvet and *cro*chet/hook), a product that consisted of two strips of nylon, one containing thousands of small hooks and the other thousands of small loops. Production began in France but moved to Manchester, New Hampshire, in 1958. Velcro's patent expired in 1978 and other manufacturers such as 3M now produce similar hook-and-loop fastening products. Velcro is used in applications as diverse as clothing, shoes, backpacks, blood pressure cuffs, sealing chambers of artificial hearts, securing gear, and clothing during space flights.

If desired, these can be explored by searching the U.S. Patent Office (*http://patft.uspto.gov*). Velcro serves as an exemplar of the Points to Ponder quotes and the field of biomimicry.

After recounting this story, discuss other seed-dispersal mechanisms such as the various means that angiosperms use insects, birds, or mammals to carry pollen or whole seeds. If available, *The Private Life of Plants* (Vol. 1, *Branching Out*), *Sexual Encounters of the Floral Kind*, *The Seedy Side of Plants,* and the free Plants-in-Motion website all provide powerful visuals with close-ups and time-lapse photography of the variety of ways that plants disperse their seeds, including the Velcro approach (see Internet Connections). These symbiotic relationships may be either a type of mutualism that benefits both species (e.g., fruit) or commensalism where the plant benefits and the animal is neither harmed nor advantaged (e.g., burrs). The University of North Carolina and Wikipedia websites and previously cited videos provide useful information. If desired, the topic of animal-plant co-evolution can be further studied by using any of the analogy-based Darwin's Finches-Pliers laboratory simulations (see Internet Connections).

In any case, be sure the learners understand how both human-engineered and natural bio-engineered "products" work in terms of the interactions of component parts to create synergistic systems where the whole is greater than the sum of the parts (i.e., emergent properties). In particular, emphasize how structures (or forms), processes, and instinctual behaviors of living creatures are all the results of millions of years of natural evolutionary bioengineering that solved problems (i.e., served particular functions) or took advantage of serendipitous opportunities to carve out unique biological niches.

Raise the following questions about the theory of evolution: Does the evolution of an array of ingenious seed-dispersal solutions support the idea of a thinking-reasoning parent plant (as suggested by the previous anthropomorphic language)? Does the form-function fit of seeds necessitate an intentional, intelligent design? If not, how do scientists account for such natural creativity and ingenuity? See the books and internet sites on pages 144 and 146–148 (and a longer list in Activity #12) for sample readings, additional activities, and multimedia resources. In grades 5–12 classrooms, several days could profitably be spent on the Explain phase.

4. *Elaborate: Bionics: Learning a Thing or Two From Nature's Engineers*

    The engineering field of *bionics* (or *biomimetics*) attempts to use solutions developed by other living organisms and natural systems as starting points for developing solutions to human problems. Dissecting and modern imaging techniques of living organisms can be viewed as *reverse engineering,* where an organism is "taken apart" to determine its basic internal design. In many cases, patented human inventions can be traced directly to natural analogs that served as the model and source of inspiration. Conversely, new human technologies are constantly helping us better study and understand (by way of analogy) how nature works.

    The two-column list on pages 143 and 144 can be used to encourage students to look at nature's design with an eye to the bionic, *form-function* connection. Possible ways to use the table include the following:

    - Scramble the order of the second column and ask students to match the columns and create a third column that describes similarities and differences between the human and natural inventions.
    - Omit the second column and ask students to research natural analogs for the human inventions (or vice versa).
    - Group the items by categories such as communication, food, housing, and protection, and assign research teams to explore the related natural adaptations and their human analogs and report back to the whole class in a report that includes multimedia, demonstrations, simulations and creative movement, and so on.
    - Use grouped items as themes for a series of bulletin board displays. Have students bring in pictures of natural and human-designed items that relate to the particular theme.
    - Have students research the conservation of plant and animal diversity from the perspective of preserving new, potentially life-saving drugs and bionic or genetic engineering solutions. The 1992 movie *Medicine Man* (starring Sean Connery) is a fictional account of one such effort.

Human Invention	Nature's Invention
	(List is not inclusive of all examples.)
air conditioning	beehives, termite mounds, and prairie dog towns
antifreeze solutions	Antarctic ice fish and the gray tree frog
armor	armadillo, hedgehog, wood louse, and turtle
autofocus and exposure lens	human eye
baby-carrying pouch	kangaroo
barbed wire	rose thorns and cactus spines
basket weaving	bird nest building
battery/electricity	electric eel (up to 600 V and 1 amp)
boat paddles	webbed feet of frogs and turtles
body language	bee dance, signaling in herds, and dominant and subordinate posturing
bridge building	spiders
camouflage	insects (Sphinx moth, Australian walking stick, etc.) and chameleon
chemical warfare	skunk, bombardier beetle, stinkbug, poison arrow frog, and snakes
chemiluminescence	firefly (a beetle) and the "cool light" of other bioluminescent species
conveyor belt	cilia of mussels and other filter feeders
dam building	beaver
division of labor and teamwork	ant, termite, and bee colonies; lichen (algae and fungi); and wolf packs
drinking straw	mosquito, butterfly, and bumblebee
evaporative cooling	ears of elephant and jackrabbit and pigs wallowing in mud
farming and gardening	leaf-cutter ants
fashion and grooming	peacocks and other birds; apes and monkeys
fiber optics	polar bear fur
fiber-reinforced	bamboo stalk, celery
flight	insects, birds, and bats (flying squirrels and gliding fish)
flipper	frogs and ducks
float	jellyfish
geodesic dome	bubbles
gliding	flying squirrel and flying fish
glue	barnacles and spiders
helicopter	elm and maple leaf seeds
high-rise apartments	termite mounds (up to 25 ft.)
hot-tub bathing	Japanese macaque monkeys
hypodermic syringe	rattlesnake fangs
infrared photography	thermoscopic heat vision of snakes that hunt mammals
jet propulsion	octopus, squid, dragonfly nymphs, and scallops
mobile homes	snails and hermit crabs
navigational systems	migrating birds, salmon, whales, and butterflies

Human Invention	Nature's Invention
	(List is not inclusive of all examples.)
net traps and webbing	spiders' webs
nylon	silk produced by silkworms
paper making	wasps
parachute	dandelion seeds
perfumes	insect pheromones and fragrant flowers
pins and needles	porcupine quills and sea urchin spines
scissors	scissor-toothed jaws and beaks of birds
showering and snorkeling	elephant's trunk
singing	humpback whales and birds
solar-tracking	sunflowers and phototropisms
sonar	echolocation of bats and various whales and dolphins
spinning projectile motion	paramecium locomotion
submarine ballast tank	swim bladder of fish and nautilus chambers
suction cup	octopus, squid, limpet (a mollusk), housefly, and tree frogs
synthetic fibers	silkworms and spiders
telephone receiver	human ear
tunneling	gopher and prairie dog
warning colors and hazard signs	bright coloration of wasps, ladybugs, and poisonous frogs
water-repelling and self-cleaning	lotus leaf of water lilies
vertical takeoff	housefly
V-flight formation	migrating geese
war	ants
zippers	barbules of bird feathers

Sources of Information

- Forbes, P. 2006. *The gecko's foot: Bio-inspiration—Engineering new materials from nature.* 1st U.S. ed. New York: W. W. Norton.

- Gates, P. 1995. *Nature got there first: Inventions inspired by nature.* New York: Kingfisher.

- Munch, T. W. 1974. *Man the engineer, nature's copycat.* Philadelphia: Westminster Press/Franklin Institute.

- Postiglione, R. A. 1993. Velcro and seed dispersal. *American Biology Teacher* 55 (1): 44–45.

• Suid, M. 1993. *How to be an inventor*. Palo Alto, CA: Monday Morning Books.

5. *Evaluate:*

   In addition to conventional paper-and-pencil summative tests, students can be asked to research and report on the form-function connection and evolutionary bioengineering as exemplified in any of the previously cited seed-dispersal mechanisms or natural "inventions."

Debriefing

When Working With Teachers

This mini-5E can lead to a discussion of the 5E cycle as a powerful multiday and unit-level planning framework (see the previously mentioned visual participatory analogy). Although individual lessons can also sometimes be designed as a mini-5E, the 5E model is most powerful when used to frame a one to three week unit, with one or more days devoted to each of the 5E phases. See the Internet Connections: BSCS and Miami Museum of Science. This particular sequence of activities can also be used to feature the AAAS *Benchmarks for Science Literacy* (1993) idea of Common Themes (chapter 11, "Systems, Models, Constancy and Change, and Scale"; see Appendix C) in teaching science. Also discuss how teachers need to challenge the common misconception that evolution is an optional, theoretical, biological topic without many significant, practical implications. In fact, evolution is both supported and used by chemistry, geology, and physics and is a broad-reaching, cross-disciplinary, unifying theme that has significant real-world applications that range from agriculture, bioengineering, and conservation to personal and global health. Teachers can read and discuss the position statements of various professional organizations on the teaching of evolution (see Internet Connections: ASPB, BSA, NABT, NCSE, and NSTA). See also activities #6, #7, and #12.

When Working With Students

This sequence of activities would make a good instructional sequence to begin a larger unit on plant reproduction or part of a yearlong

theme on evolution and the connection between form and function in nature. The Points to Ponder quotes can be discussed in light of how words such as *design, purpose,* and *function* are used in the context of the theory of evolution. Students should be encouraged to fight their own personal "nature-deficit disorder" (Louv 2008) by investing time in unplugged, audio- and WiFi-free natural settings as a source of biosensory recalibration, psycho-emotional rejuvenation, aesthetic appreciation, and intellectual inspiration.

Extensions

1. *Serendipitous Discoveries:* Electromagnetism, penicillin, Post-it notes, Scotchgard, silly putty, Slinky, Teflon, vulcanization of rubber, and x-rays are all cases where, as Louis Pasteur put it, "chance favored the prepared mind" (see Wikipedia: Serendipity for other examples).

2. *Engineering Insights:* Any of the popular books by Henry Petroski provide engaging connections between engineering and everyday products. See, for example, *The Evolution of Useful Things: How Everyday Artifacts—From Forks and Pins to Paper Clips and Zippers—Came to Be as They Are* (1994) and *Invention by Design: How Engineers Get From Thought to Thing* (1996). The latter book discusses the paper clip, wood-cased pencil, zipper, Velcro, zip-top bag, aluminum can, fax machine, and more.

3. *Close-up Perspectives:* The popular MindTrap games include a large number of close-up, magnified, "What am I?" type pictures of everyday objects with short verbal clues. Fifty-two of these have been collected in a book: Detective Shadow. 2000. *Tricky MindTrap puzzles challenge the way you see and think.* New York: Sterling Press.

Internet Connections

• American Society for Microbiology: Classroom Activities: One of These Things is Not Like the Other (textile identification and classification game using a handheld microscope): *www.asm.org/ index.php/education/classroom-activities.html*

- American Society of Plant Biologists (ASPB): Education (resources): *www.aspb.org/education*
- Biomimicry Institute (variety of resources): *www.biomimicryinstitute.org*

    In particular, download Velcro Race: *www.biomimicryinstitute.org/education/k-12/curricula.html*
- *Bioteams* (book by Ken Thompson applies nature's wisdom to human teams and organizations): *www.bioteams.com*
- Botanical Society of America (BSA): Statement on Evolution (see also other teaching resources): *www.botany.org/outreach/evolution.php*
- *BSCS 5E Instructional Model: Origins, Effectiveness and Applications:*

    www.bscs.org/pdf/5EFull Report.pdf (65 pages)

    http://bscs.org/pdf/bscs5eexecsummary.pdf (19 pages)
- Darwin's Finches and Pliers Laboratory Simulation:

    www.scribd.com/doc/13189835/Galapagos-Finch-Lab

    www.accessexcellence.org/AE/AEC/AEF/1996/sprague_beaks.php

    http://biol1114.okstate.edu/study_guides/labs/lab6/lab6_frameset.htm

    http://biochemnetwork.com/download.php?id=113

    www.sciencecases.org/darwins_finches/prelude.asp (case study with PowerPoint slides)
- Miami Museum of Science: Constructivism and the 5E: *www.miamisci.org/ph/lpintro5e.html*
- National Association of Biology Teachers: NABT Position Statement on Teaching Evolution: *www.nabt.org/websites/institution/index.php?p=35*
- National Center for Science Education (NCSE): Defending the Teaching of Evolution in Public Schools: *http://ncse.com*
- National Science Teachers Association (NSTA): Position Statements:

    The Teaching of Evolution: *www.nsta.org/about/positions/evolution.aspx*

- Science and Plants for Schools (teaching resources): *http://www-saps.plantsci.cam.ac.uk*

- Teachers' Domain:

    Design Inspired by Nature (lesson plan and slide show): *www.teachersdomain.org/resource/eng06.sci.engin.design.biomimicry*

    The Remarkable Cocklebur: Worldwide Hitchhiker and Nature's Velcro (essay with photos): *www.teachersdomain.org/resource/tdc02.sci.life.stru.cocklebur*

- University of NC Learning/NC Lesson Plans: Modify a Seed Activity (model seed dispersal with a dry kidney bean): *www.learnnc.org/lessons/BertWartski5232002651*

- U.S. Patent and Trademark Office: Velcro, Patent No. 2717437: *http://patimg1.uspto.gov/.piw?Docid=2717437&idkey=NONE*

- Velcro Invention: George de Mestral: *http://inventors.about.com/library/weekly/aa091297.htm*

- Videos and DVDs that feature the variety of strategies used by plant seeds to aid in their dispersal:

    Carolina Biological: *Sexual Encounters of the Floral Kind*: DVD tape (60 min.; 495932D; $79.95): *www.carolina.com*; 800-334-5551

    David Attenborough's *The Private Life of Plants*: 2 DVDs (300 min.; $39.95): *http://shop.abc.net.au/browse/product.asp?productid=724292* (The BBC series is described at *http://en.wikipedia.org/wiki/The_Private_Life_of_Plants.*)

    Nature (PBS series): *The Seedy Side of Plants*: *www.pbs.org/wnet/nature/plants*

    Plants-In-Motion (free time-lapse photography QuickTime movies): *http://plantsinmotion.bio.indiana.edu*

- Wayne's Word: *An Online Textbook of Natural History*: Noteworthy Plants (more than 40 articles; e.g., Remarkable Cocklebur: Nature's Velcro): *http://waynesword.palomar.edu/index.htm*

- Wikipedia: *http://en.wikipedia.org/wiki*. Search topics: bidens (or sticktight), biological dispersal (seeds), biomimicry, bionics, burdock, cocklebur, greater burdock, serendipity (links to large number of examples), symbiosis, and Velcro.

Answers to Questions in Procedure, steps #1–#3

1. *Engage: Clothing Conundrum:* Clothing can be attached in a variety of ways: ropes and belts, buttons (1200s), safety pins (antecedents back to 1000 BC, but the "safety" version was developed in 1849), snaps (1885), zippers (1893), and Velcro (1955).

2. *Explore:*

    a. *Clothing Close-up: Views of Velcro:* A hook-and-loop design form or structure underlies the functional operation of Velcro. The handheld microscope makes this design very easy to see.

    b. *Plant Parenting Problem:*

    (1) If seeds are germinated too close to a parent, they would compete with the parent for space, soil nutrients, water, and sunlight. This would also be undesirable for the seed that would be overshadowed by the parent plant, both literally and figuratively. Wider dispersal of offspring also increases the odds of species survival and the evolution of new adaptive modifications for new environments.

    (2) Though some plants, such as tumbleweeds and resurrection plants (see Activity #6), can move from place to place, most seeds are transported away from the parent plant that does not relocate. Seeds may be packaged in attractive-smelling, tasty, nutritious fruit or nuts that will be ingested and subsequently deposited in animal waste; some float on water currents, such as coconuts; some use "parachutes" such as dandelions and milkweed; some use "helicopters" such as maple seeds; and so on.

    (3) Burdock seeds "hitchhike" by sticking to animal fur in a manner analogous to Velcro.

    (4) The two systems are dissimilar in that plants only needed to invent "half" of the mechanism, and animal fur (originally designed for a different purpose) provided the other half.

3. *Explain: Science Stories: Making Connections:* Seeds that are spread unwittingly by animals represent examples of co-evolution where either both species benefit from the arrangement (i.e., analogous to animal pollinators that assist plants' sexual reproduction or the mutually beneficial nature of fruits and nuts that provide nutrition for animals and seed transportation for plants) or the plant is the sole beneficiary. Plants lack a nervous system to enable them to strategically plan "intelligent" solutions to such species' survival challenges. Instead, genetic variation, mutation, and natural selection over millions of years lead to survival-of-the-fittest designs (i.e., relative to particular environments).

Activity 10

Osmosis and "Naked" Eggs
The Environment Matters

How will I be able to hold myself together when my shell is gone?

VINEGAR

Expected Outcome

Fresh, uncooked, de-shelled eggs (i.e., eggs previously placed in an acid bath) are observed to increase in volume when placed in water or to decrease in size when placed in corn syrup. This macroscopic model of osmosis can be used as an introduction to the evolution of eukaryotes and multicellularity as a means to grow beyond the size limitations of very small, microscopic prokaryotic cells.

Science Concepts

Osmosis is the diffusion of water through a selectively permeable membrane. It is a form of passive cellular transport (i.e., it doesn't "cost" the cell energy) where water moves from a region of higher water concentration to a region of lower water concentration (until a dynamic equilibrium is reached when the water concentrations are equal). Unfertilized, haploid mammal and bird eggs (or just-fertilized, diploid ones before they start to divide and differentiate) are unusually large, macroscopic single cells. The human egg is about the size of a period and bird eggs can be as large as an ostrich's egg; in both cases, the embryo itself is initially a very small part of the egg's total volume, which is primarily food. An egg's shell must be able to contain the vital liquid water content of the egg, yet porous enough to allow oxygen used in the growing embryo's metabolic processes to enter and to release the carbon dioxide that is produced as a result. A soft, flexible cell membrane lies just beneath the hard exterior shell of the egg. Given their size and availability, unfertilized chicken eggs provide a convenient macro-*scale model* of the *system*-level phenomenon of osmosis (as driven by invisible molecules-in-motion).

Once the shell of a fresh, uncooked chicken egg is removed in an acid bath, the egg's intact, selectively permeable cell membrane is elastic enough to noticeably swell and shrink without bursting in response to changing osmotic pressure. Random movements of relatively smaller, faster-moving water molecules cause the net flow of water to move from the side of the membrane where they are more concentrated to the side where they are less concentrated. Egg white is about 90% water and corn syrup is about 25% water. Thus, if a de-shelled egg is placed in corn syrup, water moves from inside the egg to outside the egg, leaving the egg shrunken and limp (versus the de-shelled egg placed in pure water, which swells and tightens). As such, the de-shelled egg-solution system demonstrates how *constancy and change* (i.e., dynamic equilibrium) define biological systems.

Students are likely to have a variety of misconceptions and conceptual "holes" related to this mini unit, including the following:

1. All cells are microscopic.

2. Macroscopic, unfertilized, grocery store chicken eggs were either never alive or are made of many now-dead cells that do not display any of the characteristic processes of living cells.

3. Shells in bird eggs and cell walls in bacteria and plant cells are substitutes or replacements for cell membranes (which unfortunately sometimes are not clearly highlighted on textbook visuals).

4. Diffusion-related concentration gradients always cause the movement of the solute, not the solvent. Students are also unlikely to differentiate osmosis from active transport that enables cells to preferentially maintain higher unique combinations of various biomolecules and ionic concentrations inside (versus outside) the cell. This capability is what distinguishes living cells from their external environments. Clarifying these misconceptions also allows osmosis to be used as a lead-in to an Elaboration-phase analysis of cell surface-area-to-volume restrictions and the evolution of membrane-bound cellular organelles and multicellularity (in eukaryotes) from simpler, single-celled prokaryotes.

If desired, the corrosive effect of an acid on calcium carbonate can also be discussed as a visual analogy for acid rain's effect on marble and limestone, which are also forms of $CaCO_3$.

Science Education Concepts

As a *visual participatory analogy*, students' minds can be viewed as selectively permeable membrane-bound systems that allow certain inputs and outputs (and reject or restrict others) that vary with the particular curriculum-instruction-assessment (CIA) environment in which the students are placed. Intelligently designed CIA environments facilitate a "loss" of misconceptions, a "gain" of scientifically valid conceptions, and an overall expansion of understanding and capacity for future learning. Poorly designed CIA environments cause students' minds to "shrink" in capacity for future learning (i.e., are mis-educative). Of course, a critical limitation of this analogy is that learning is not a form of passive transport or "absorption" of static external facts, but the learner-active, energy-requiring construction of new and reconstruction of old conceptual

schemas in light of new experiences via assimilation and accommodation. 5E Teaching Cycles (Engage, Explore, Explain, Elaborate, Evaluate) are designed to scaffold learning across a planned, logical, dynamic sequence of student-teacher, student-student, and student-phenomena interactions.

Materials

- For a teacher demonstration, 4 fresh, uncooked eggs are needed. If done as a hands-on exploration, a dozen (12) eggs for a class would allow 2 eggs for each of 6 groups of 4 students.
- For each egg, you will need about 250 ml of vinegar (dilute acetic acid), 250 ml of colorless corn syrup, and a 500 ml glass beaker with a piece of aluminum foil and rubber band or plastic wrap to control odor (or alternatively, 16–18 oz. plastic, sealable jars).
- Measuring tape (or string and a metric ruler) and an optional balance are needed to measure pre- to postosmotic changes in size and mass.
- A colander or screen serves as a macroscopic model that allows small particles (e.g., sand or BBs) to pass through, but not larger ones (marbles). The use of a plastic zip-top bag, tincture of iodine solution, and (corn)starch solution serves as another model system.
- Several boxes of sugar cubes or a box of 1 cm³ mathematics cubes (for a HOE) and/or several shoe boxes (for a demonstration) are also needed.
- Cubes of raw potato or gelatin, acid-base indicators, vinegar or dilute (0.1–0.4%) sodium hydroxide, and a soap bubble solution are needed for optional elaboration phase activities.

Time note: For use as a demonstration when working with teachers, prepare de-shelled eggs and then immerse them in water or corn syrup solutions for several days prior to class. About 20 minutes of discussion time is needed to quickly take teachers through the first three phases of a modified 5E Teaching Cycle. A variety of biological demonstrations require this TV cooking-show-type advance preparation (e.g., Bilash and Shields 2001; Ingram 2003; VanCleave 1989, 1990, 1991a, 1991b).

Points to Ponder

Contact with things and laboratory exercises, while a great improvement upon textbooks arranged upon the deductive plan, do not of themselves suffice to meet the need. While they are an indispensable portion of the scientific method, they do not as a matter of course constitute scientific method. Physical materials may be manipulated with scientific apparatus, but the materials may be disassociated in themselves and in the ways in which they are handled, from the materials and processes used out of school ... There is sometimes a ritual of laboratory instruction as of heathen religion.

—John Dewey, American philosopher-educator (1859–1952) in his 1916 book *Democracy and Education* (see Internet Connections)

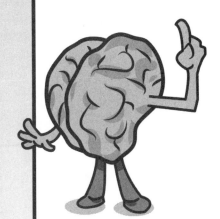

... learning is a dual process in which, initially the inside beliefs and understandings must come out, and only then can something outside get in. It is not that prior knowledge must be expelled to make room for its successors. Instead these two processes—the inside-out and the outside-in movements of knowledge—alternate almost endlessly. To prompt learning, you've got to begin the process of going from the inside out. The first influence on new learning is not what teachers do pedagogically but the learning that's already inside the learner.

—Lee Shulman, "Taking Learning Seriously," *Change* 31 (4): 10–17.

Procedure

This 5E sequence can be completed as a demonstration-experiment (with teachers in a cut-time frame) or as a lengthier series of hands-on exploration (with grades 5–12 students).

Engage: Macroscopic Models and Egg-citement About Cells

1. If used as a Teacher Demonstration: Between two and four days prior to beginning the activity, use a tape measure to determine the vertical and horizontal "circumferences" of four fresh, uncooked eggs. Submerge the eggs in vinegar. The weak, dilute acetic acid in the vinegar (i.e., approximately 5% acid and 95% water) will chemically react with the calcium carbonate in the egg's shell to form water-soluble calcium acetate, carbon dioxide gas (bubbles), and water. If the shells are not completely removed in 24 hours, replace the used, reacted vinegar with fresh vinegar and let the eggs soak for another 24 hours. After the shells are completely gone, rinse two of the three de-shelled, translucent eggs with water, re-measure their circumferences, weigh them, and submerge one in colorless corn syrup and the other in water (leave the third in the vinegar). Within a day, the first egg will lose water to the corn syrup and dramatically shrink in size and mass, and the second one will gain water from its water bath (the third left in vinegar should remain fairly constant). The changes can be measured and tabulated over a one to four day period. Be sure to cover the beakers or use sealable jars to control odor.

2. On the day of the demonstration, introduce this mini unit by briefly highlighting some key historical events. In 1665, Robert Hooke, an English scientist who developed some of the earliest compound microscopes, coined the term *cells* to describe the walled, empty chambers he observed in slices of cork at 30× magnification that reminded him of the cells in monasteries. His contemporary, a Dutchman amateur scientist named Antonie van Leeuwenhoek, was simultaneously discovering a wide variety of what we now know to be living, motile, single-celled organisms everywhere he looked with his 10-times-more-powerful single-lens microscopes. Improvements in microscopy led to discoveries of increasingly varied and ubiquitous

cells in and on macroscopic living creatures and on land and in water. Nearly two centuries later, in the late 1830s, a German physiologist-anatomist (Theodore Schwann) and botanist (Matthias Schleiden) both wondered if cells were the biological equivalent of chemists' atoms (as proposed by John Dalton in 1808). That is, were cells the basic unit of all life? Let's explore this and related questions:

a. Are all living things composed of cells? Are all cells microscopic? Hold up a fresh, raw egg and ask: Can a chicken egg bought in a grocery store be considered a single large cell?

b. What is the most basic and universal component of every living cell (from the smallest and simplest prokaryotic to the largest or most complex eukaryotic cell) and what is this structure's function?

c. Do chicken eggs have a cell membrane? Is it the same as the shell?

d. Is there a way to remove the shell of a fresh, uncooked egg (without hard-boiling it!) and still have the egg maintain its shape and contents? What is the main chemical ingredient of eggshells? What liquids might chemically react with and dissolve this chemical compound?

Optional Teacher Demonstrations: Place a fresh egg, eggshell, or a piece of white chalk in a hardware-store-bought solution of muriatic acid (HCl sold as a masonry cleaner) or laboratory hydrochloric acid (the more concentrated, the faster the reaction). This is much quicker and more dramatic than using the more dilute and weak acetic acid found in vinegar. Alternatively, you may wish to contrast the effectiveness and speed at which "Easter egg" color-dyeing occurs with versus without the recommended vinegar solution. In either case, the demonstrations can be reserved for a postlab activity that leads into the STS issue of acid rain.

Safety Note

Wear indirectly vented chemical-splash goggles.

e. *If the activity is to be done as an HOE:* Have learner teams measure the circumferences of their two raw eggs and place them in vinegar in sealed jars for a lab that will resume in a day or two after the de-shelling. If done as a demonstration, proceed directly to the Explore phase.

Explore: Eggs-perimental Science: If used as a Teacher Demonstration:

3. a. Tell the learners you have previously removed the egg-shells from two eggs in a similar (but slower) fashion than in step #2d and that you would like them to examine the results. Distribute for up-close inspection (or use a projection video camera to display the image of) two de-shelled, "naked" eggs that have been soaking in vinegar. If disposable plastic gloves or soap and water are available for subsequent hand cleaning, allow the learners to hold a de-shelled egg in their hands to feel the osmotic pressure. Alternatively, the de-shelled eggs can be placed in zip-top bags before allowing the students to handle them without risking the mess of a ruptured egg. Also, the inside yolk can be more readily noticed when the room is darkened and a flashlight beam is held against the egg. Draw attention to the thin, translucent, flexible cell membrane that is still keeping the egg's contents inside. Ask the learners: What everyday material does the cell membrane reminds you of? Do you think certain kinds of materials could pass through the cell membrane without rupturing it? How might you test if a cell membrane is "permeable"? What kind of materials would need to pass through an egg's cell membrane (and the egg itself) if the egg had been fertilized and contained a developing embryo? Why?

   b. Display or circulate the two other de-shelled eggs. As they examine the one that has been soaking in water and the other in corn syrup for 1–3 days, ask: What is the main chemical component of the interior of eggs? How could we test whether movement of water is occurring through the membrane?

   c. Lead the learners to see that the egg's cell membrane allowed water to enter into the egg in the case of the water bath and water to leave the egg in the corn syrup solution.

   d. *Optional Teacher Demonstration*: The Steve Spangler Science website (How to Make a Folding Egg) describes a classic magician trick that relies on an "empty" egg cell membrane made by poking two small holes on opposite ends of a fresh, clean egg and shaking the egg to "scramble" the contents, which are then removed by blowing on one end of the egg before soaking it in an acid to remove the shell. The empty cell membrane can be used to demonstrate the thinness of the cell membrane.

If the activity is to be done as an HOE, steps #3 a–c would be developed collaboratively with students as a part of a discussion on experimental design.

Explain: Just Passing Through: Modeling the (w)Hole Truth About Osmosis

After the data have been tabulated and the learners have been challenged to use logical arguments and skeptical review to explain the empirical results, demonstrate the following two macroscopic models of the molecular level process.

Part 1: Colanders Count (and Sort by Size): Use a colander or screen as a physical model to demonstrate a semipermeable or selectively permeable membrane that "selects" for small molecules by "sifting and sorting" a mixture of marbles and BBs or BBs and sand. Ask:

4. a. If the colander or screen represents the cell membrane, what property of molecules is it "selecting for"?

    b. Which size particles in this model represent water and which ones represent sugars, proteins, and other biomolecular components of the cell?

    c. Because the smaller particles can pass either way through the colander, what factor determines which way they move in this model? How is this different than a real cell membrane?

    d. Rather than a gravity gradient, what factors determines the direction of water molecules in our egg cell model? Why is two-way movement important for cells?

Part 2: Plastic Baggie Membrane Model: Demonstrate that drugstore tincture of iodine (I_2 dissolved in alcohol) or Lugol's solution (I_2 dissolved in an aqueous solution of KI) turns a dark blue-purple in the presence of starch and thus serves as a chemical indicator for this large polysaccharide (but not for mono- or disaccharides). Use a zip-top bag as a model of the cell membrane. Place a starch solution (e.g., cornstarch, diced potatoes, or eco-foam

or starch packing peanuts dissolved in water) on the inside of the bag, being careful to not spill any on the outside. Then push out any air, seal the bag, and rinse the outside with water. Place the sealed bag into a beaker that contains a dilute tincture of iodine or an I_2-KI solution. Prepare a comparable setup with the location of the two solutions reversed. Ask students to predict-observe-explain what will occur in the two setups. Challenge them to make drawings that depict relative sizes and concentrations of the various components of this system prior to showing dynamic macroscopic simulations and/or computer animations (see below). *Note:* I_2 is a comparatively small molecule and starch is a quite large molecule, $(C_6H_{10}O_5)_n$. The relatively large starch molecules cannot diffuse through the smaller intermolecular "holes" in the plastic membrane, but the smaller iodine molecules can. In *Brain-Powered Science*, Activity #20, Extensions #2 and #3 (pp. 217–218) describe a soap bubble–cell membrane analogy and the diffusion of vanilla through a sealed, inflated balloon; see also the Access Excellence and Nanopedia websites in Internet Connections.

Part 3: Simulating the Sorting Sequence: See the Internet Connections for a number of excellent molecular simulations on diffusion and osmosis. Computer animations and/or people-as-molecules simulations are powerful means of helping students visualize molecular motions and mechanisms that are otherwise unseen and difficult for students to appreciate. Learners can also explore other systems that demonstrate osmosis, such as those that use dialysis tubing.

Reading selections from the textbook and internet sources (especially those with animations) can be used to supplement the class discussions. Be sure to feature the "big ideas" of systems, models, constancy and change, and scale (see Appendix C). Also, live, projected microscopy or multimedia clips can be used to challenge students to predict-observe-explain what happens to living cells when they are placed in hypertonic, isotonic, and hypotonic solutions.

Elaborate:

Part 1: How Sweet It Is to Have a Large Surface-Area-to-Volume Ratio:
Return to these questions: Why are nearly all cells limited in size to the microscopic realm? Why don't we observe lots of macroscopic unicellular organisms (as commonly featured in science cartoons; see the Science Cartoons Plus website in Internet Connections)? How does the size of a cell influence its ability to exchange materials (e.g., water, nutrients, and waste products) with its environment through osmosis (as well as via active transport)? Despite textbook pictures and views through standard microscopes that suggest otherwise, all cells are 3D, not 2D. As a cell increases in size (volume, or V), does it increase proportionally in surface area (SA)?

Teacher Demonstration

Use identical shoe boxes to construct model cells that are made up of one, two, three, and four boxes. Make a quick qualitative argument about the loss of effective (exposed) surface-area-per-unit volume when the boxes are stacked on top of (and/or alongside) each other.

Student Hands-on Explorations

Distribute variable numbers of 1 cm^3 math blocks (or use sugar cubes as a somewhat larger substitute) to teams of students to construct 3D cubes of increasing linear dimensions (i.e., 1 cm × 1 cm × 1 cm [1 block], 2 cm × 2 cm × 2 cm [8 blocks], 5 cm × 5 cm × 5 cm [125 blocks]). Have each team build and calculate the surface area (SA) and volume (V) for two different-size cubes. *Note:* For a cube, SA = $6s^2$ and V = s^3 where s = length of a side. Alternatively, rather than using blocks as cells, use water-filled zip-top bags. Students can measure the surface areas of different-size zip-top bags (up to the seal), then measure what volume of water is needed to fill them. In either case, collectively tabulate and share the data to reveal that the SA/V ratio (= $6s^2/s^3$ = $6/s$ for cubes) decreases dramatically as the overall size increases because surface area increases by the square and volume by the cube of the linear dimension. Cells that grow "too large" would suffer from limited access to nutrients and oxygen and an inability to rid themselves of their own metabolic waste products! Similarly, because mass is proportional to volume for a given substance with a given density (e.g., 1 cm^3 of water weighs 1 g), the mass

of the cubes also increases faster than the surface area, which makes issues of structural support a problem for nonaquatic organisms.

If desired, challenge the learners to develop an experiment that uses biological material to empirically verify the logic of the mathematics of surface-area-to-volume ratios. *Idea starter:* A raw potato can be cut into cubes of increasing size, submerged in a dilute tincture of iodine solution for a period of time, and subsequently dissected to compare how far into the different potato cubes the iodine solution (an indicator that turns a dark blue-purple in the presence of starch) penetrated. Alternatively, different-size agar or uncolored gelatin cubes can be made with an acid-base indicator (e.g., phenolphthalein or bromthymol blue) dissolved in them. When placed in the appropriate solution (e.g., 0.1–0.4% sodium hydroxide or vinegar, respectively), the solution will diffuse into the cube's interiors (and effect a color change) at a depth related to the surface-area-to-volume ratio of the cubes. Versions of these labs are sold by science supply companies and appear on the web. Again, online simulations (Internet Connections) can be used to complement the macroscopic models.

Part 2: Life Invents Bigger Cells and Multicellularity to Move From the Micro to Macro Worlds: Typical prokaryotic cells (bacteria) are 1–10 × 10^{-6} m in length (or diameter), as compared to typical eukaryotic cells that are 10–100 × 10^{-6} m. These differences in linear dimensions of 10–100 translate to factors of 1,000–1,000,000 in volume differences between the two types of cells (see *More Brain-Powered Science*, Activity #17, Extension #1). This raises questions such as the following:

- Why are nearly all cells microscopic?
- How do eukaryotic cells bypass the limits that seem to restrict bacterial cell size?
- What did nature "invent" to get around the size limits of unicellular eukaryotes?

Optional Soap Bubble Membrane Model: Commercial or homemade soap bubble solutions can be used to model the membrane-bound organelles of eukaryotes. Blow a large semispherical bubble on a wet sheet of overhead transparency; dip a straw into the bubble solution;

and insert the straw through the first bubble to blow a bubble within a bubble (to represent a membrane-bound nucleus) and withdraw the straw. It is also worth noting that a sphere is the 3D shape that has the smallest surface-area-to-volume ratio for a given volume. That is, it requires the least amount of material to "build" an enclosed space to contain a certain volume of material. Have the learners calculate the surface area of a square ($6s^2$) versus a sphere ($4\pi r^2$) of the same (arbitrary) volume and discuss why spherical cells would make "economic sense" for cells. Scale effects are important to life at all levels of organization. See Internet Connections: Access Excellence and/or *Brain-Powered Science*, Activity #20, Extension #2.

Evaluate:

Learners could be challenged to apply and explain the science of osmosis as related to real-world applications such as the following:

a. Few aquatic species are adapted to inhabit both freshwater and saltwater environments.

b. Contact lens cleaning solutions and intravenous (IV) solutions need to be isotonic with the patient's eye or bloodstream.

c. Shipwrecked sailors should not drink salt water because it is hypertonic relative to their cells.

d. Road salt used for de-icing poses a problem for plants and animals that live in the soil near the road.

e. Fresh vegetables can be kept crisper in grocery stores if they are periodically sprayed with a water mist or, when taken home, soaked in water or placed in a more moist section of a refrigerator or in sealed plastic bags.

Learners could be asked to predict-observe-explain the outcomes of any number of other demonstrations and experiments described in various books and biology education websites that relate to osmosis and/or surface-area-to-volume scale effects (e.g., the functioning of kidney, the small intestines, and the lungs). See Internet Connections as a starting point.

Debriefing

When Working With Teachers

Discuss the positive and negative attributes of the de-shelled egg as a visual participatory analogy for a student learner and how it relates to Shulman's quote with its emphasis on the dynamic interaction between "the inside and outside" (see also *Brain-Powered Science*, Activities #11, "Super-Absorbant Polymers," and #20, "Needle Through the Balloon"). Clearly, the CIA environment does make a difference in terms of student cognitive gains! The 5E Teaching Cycle offers multiple and different kinds of learning opportunities for students to build a solid conceptual understanding of a given topic. Also, challenge the teachers to consider the relevance of John Dewey's assessment of conventional laboratory "exercises" in the Points to Ponder quotes. Teachers may want to read and discuss a more recent critique of standard laboratory "exercises" (Singer, Hilton, and Schweingruber 2006; see Internet Connections: *America's Lab Report*).

As time permits and need requires, review the science of osmosis and discuss how the demonstration version could be modified for a direct hands-on exploration (HOE) (e.g., quantitative measurements of the changing mass and circumference of the eggs over a three to five day time period, including attempts to reverse the process for both swollen and shrunken eggs). Several versions of related labs and demonstrations exist on the internet.

When Working With Students

When used as either a demonstration and/or an HOE, this 5E mini unit fits into a larger unit on cells. A key factor in explaining osmosis is getting students to focus on the concentration of the solvent, water (in a water solution), when they more commonly focus on the opposite, the concentration of solutes in a water solution. Diffusion always involves movement down a concentration gradient. In cells, water, oxygen, carbon dioxide, and glucose molecules are all small enough to diffuse through the intermolecular "holes" in cell membranes. A central task of every cell is to maintain homeostasis or internal *constancy* amidst external *changes* in its aqueous environment that has a very different overall chemical composition. The fact that the acid bath "eats" the hard eggshell but not the cell membrane can be used to emphasize the "tough but thin" nature of the cell membrane (though, in fact, most living cells would not survive such an acid bath, even if their cell membranes did).

Osmosis is only part of the story. Some single-celled organisms (e.g., *Paramecium*; see the Internet Connections for Activity #7) have specialized structures called water vacuoles to help with water balance. Also, cells are constantly exchanging other chemicals with their environments. The movement of these other chemicals (e.g., larger molecules and ions) involves active transport that requires cellular energy to work against simple, concentration-based diffusion gradients. The eukaryotic cell's invention of membrane-bound organelles can be introduced as a means of increasing surface-area-to-volume ratios for performing chemical reactions and exchanges within the cell. Bacteria, algae, fungi, and plants have their cell membranes within an outer, rigid cell wall that maintains the shape of the cell, provides physical protection, and prevents the cell from bursting in a hypotonic environment. The latter can be modeled by blowing up a balloon (cell membrane) within the foot of an old nylon stocking (cell wall). The stocking will cause the balloon to resist further expansion when "it is full." One way that antibiotics kill bacteria (but not viruses) is by preventing bacteria from forming functional cell walls. Cell walls do not protect against plasmolysis in a hypertonic solution; this explains the antibacterial properties of salted meats and the problem plants have with road salt (see Internet Connections: DarylScience).

If desired, the first part of the demonstration (i.e., the acid bath) can also be used in the Engage phase of a unit on the effects of acid rain on limestone and marble statues and buildings, as well as on wildlife. Rainwater is naturally slightly acidic due to the normal presence of carbon dioxide in our atmosphere. But industrial activity has caused pH levels to drop lower than 5 due to the large-scale burning of fossil fuels that produce carbon dioxide, sulfur oxides, and nitrogen oxides that combine with atmospheric moisture to produce "acid rain" that threatens plants and animals and increases corrosion of metals and buildings.

Extensions

See the Elaboration phase and the Internet Connections for a variety of biological activities related to osmosis. Alternatively, when working with teachers, emphasize the amazing number of science concepts that can be explored using low-cost, everyday materials such as eggs in FUNdaMENTAL ways. Consider the following Egg-citing Physical Science Egg-speriments:

1. *Egg in the Bottle Demonstrations:*
   a. The biology of osmosis can be used to reduce the size of an uncooked, de-shelled egg by first placing it in corn syrup; then, after the egg has shrunk, place it in the bottle; pour distilled water into the bottle to "grow" the shrunken egg back to its original size or larger; and, finally, pour the water out of the bottle to leave an enlarged egg in the bottle. The last step should be done immediately before the class enters the room.
   b. The physics or chemistry gas-law-related alternatives are much quicker. Place a de-shelled hard-boiled egg on a bottle that has just been filled halfway with hot water, shaken, then emptied. The mouth of the bottle should be just slightly smaller than the egg's diameter. As the hot air trapped in the "empty" bottle cools, the internal air pressure will drop relative to the external atmospheric pressure, which forces the egg into the bottle. An even quicker way of getting the egg into the bottle is to drop a burning piece of paper into an "empty" bottle, then place the de-shelled egg on top. Challenge students to use prior knowledge to suggest ways of removing the intact egg from the bottle (e.g., heat the bottom of an upside-down bottle). See also the Internet Connections: Steve Spangler Science for a variation that uses a water-filled balloon.

2. *Egg-on-End Magic Challenge*: Challenge students to stand a fresh, uncooked egg (in its shell) upright on its end. This can be done in various ways. Prior to your public (teacher-only) demonstration of this feat, repeatedly drop muriatic (HCl or hydrochloric) acid on the big end of the egg to dissolve just a small portion of the eggshell, exposing a soft area of the egg membrane. Your pretreated

166

egg will stand up on this presoftened end. Alternatively, vigorously shake a fresh egg to break the egg yolk "free from its moorings." When this slightly scrambled egg is placed on its end, the heavier yolk will sink down within the egg white, lowering the egg's center of gravity and stabilizing it. Without the benefit of a "trick," students will not be able to replicate this feat. The difference between magic and science is understanding the "why," and a teacher's goal is to promote science without destroying students' sense of wonder about the real magic of nature.

3. *Eggs-traordinary Fluoride Protection and Tough Teeth:* Smear fluoride toothpaste on the entire surface of an egg; let it sit for 24 hours, then wash it off with warm water. Compare the effects of a vinegar bath on this treated egg versus an untreated one. The fluoride ions react with the calcium ions in the shell and provide an acid-protective coating similar to the effect of protecting tooth enamel from acidic saliva and foods. This simple activity demonstrates how knowledge of biochemistry allows humans to live longer, healthier lives than our ancestors (who suffered greatly during their short lives from premature tooth decay and a host of other problems that modern science has solved).

4. *More Brain-Powered Science*, Activity #3, "Dual-Density Discrepancies," Extension #1: Egg-citing Egg-speriment

5. *Brain-Powered Science*, Activity #29, "Rattlebacks," explores egg spinning and rotational dynamics.

Internet Connections

• Access Excellence:

Investigating Cell Sizes: *www.accessexcellence.org/AE/ATG/ data/released/0239-AltonBiggs/index.html*

The Cell Membrane and Surface Area Demos: *www. accessexcellence.org/AE/ATG/data/released/0307- TrumanHoltzclaw/index.html*

Modeling Limits to Cell Size: *www.accessexcellence.org/AE/ AEC/AEF/1996/deaver_cell.html*

Egg Osmometers: Teacher Notes: *www.accessexcellence.org/AE/ATG/data/released/0519-NancyIversen/index.html*

Diffusion and Osmosis: *www.accessexcellence.org/AE/ATG/data/released/0081-JeffLukens/description.html*

Using Bubbles to Explore Membranes: *www.accessexcellence.org/AE/AEC/AEF/1995/wardell_membranes.php*

• *America's Lab Report: Investigations in High School Science*: *www.nap.edu/catalog/11311.html*

• *BSCS 5E Instructional Model: Origins, Effectiveness and Applications*: *www.bscs.org/pdf/5EFull Report.pdf* (65 pages)

• Biology Corner: Cells (diffusion, osmosis, and more): *www.biologycorner.com/lesson-plans/cells*

• Cell Biology Animations: *www.johnkyrk.com*

• Cell-ebration Homepage: *www.usd.edu/%7Ebgoodman/Cell-ebrationframes.htm*

• Concord Consortium: free downloadable simulations: *http://mw.concord.org/modeler*

Diffusion, Osmosis and Active Transport: *www.concord.org/activities/diffusion-osmosis-and-active-transport*

Osmosis:

www.concord.org/~btinker/workbench_web/models/osmosis.swf

http://mw.concord.org/modeler/molecular.html

• Cornell Institute of Biology Teachers: Physiology: Diffusion Across Biological Membranes: *http://cibt.bio.cornell.edu/labs/phys.las*

• DarylScience: Biology Demos: *www.darylscience.com/DemoBio.html*

See demonstrations #7, "Crisp or Limp Salad"; #8, "Exploding Blood Cells"; #11, "How Does Penicillin Work?"; #12, "A Bill Nye Quickie (Vanilla and Balloon)"; see also *Brain-Powered Science*, Activity #20, Extensions.

- *Democracy and Education*: 1916 book by John Dewey: Columbia University's Institute for Learning Technologies' Digital Text Project: *www.ilt.columbia.edu/Publications/dewey.html*

- Disney Educational Productions: *Bill Nye the Science Guy*: Cells ($29.99/26 min. DVD): *http://dep.disney.go.com*

- Egg-citing Egg-speriment: *www.haverford.edu/educ/knight-booklet/theegg.htm*

- Exploratorium: *www.exploratorium.edu/cooking/eggs/kitchenlab.html*

- HyperPhysics, Department of Physics and Astronomy, Georgia State University: Fluids: select diffusion, osmosis and membrane transport for concept maps and explanations: *http://hyperphysics.phy-astr.gsu.edu/hbase/hframe.html*

- Miami Museum of Science: Constructivism and 5E: *www.miamisci.org/ph/lpintro5e.html*

- Minnesota Science Teacher Education Project: Osmosis and Gummy Bears: *http://serc.carleton.edu/sp/mnstep/activities/26990.html*

- Nanopedia: Soap Bubbles as Nanoscience (with embedded link to cell membranes): *http://nanopedia.case.edu/NWPage.php?page=soap.bubbles*

- National Center for Case Study Teaching in Science: Osmosis Is Serious Business: *http://sciencecases.lib.buffalo.edu/cs/collection/detail.asp?case_id=283&id=283*

- Polymer Science Learning Center: Diffusion and Gummy Bears: *http://pslc.ws/macrog/kidsmac/activity/bear.htm*

- Salmonella Poisoning:

    Centers for Disease Control and Prevention: *www.cdc.gov/salmonella*

    Mayo Clinic: *www.mayoclinic.com/health/salmonella/DS00926/DSECTION=symptoms*

    Medicine.com: Salmonella: *www.medicinenet.com/salmonella/article.htm*

- San Diego State University: Biology Lessons for Prospective and Practicing Teachers:

  Cells: *www.biologylessons.sdsu.edu/ta/classes/lab7/index.html*

  Osmosis: *www.biologylessons.sdsu.edu/ta/classes/lab5/index.html*

- Science Cartoons Plus: The Cartoons of S. Harris (e.g., examples of macroscopic cells): *www.sciencecartoonsplus.com/index.php*

  See also other sites listed in Activity #12.

- Science-Class.net: Osmosis and Diffusion (contains many external links for middle-school level): *http://science-class.net/Biology/Osmosis.htm*

- Science NetLinks AAAS: Plasmolysis in Elodea Plant Cells (lab): *www.sciencenetlinks.com/lessons.cfm?BenchmarkID=5&DocID=106*

- Serendip: Hands-on Activities for Teaching Biology to High School or Middle School Learners: See Diffusion and Investigating Osmosis: *http://serendip.brynmawr.edu/sci_edu/waldron*

- Steve Spangler Science: The Egg in the Bottle Trick and How to Make a Folding Egg:

  www.stevespanglerscience.com/content/experiment/00000022

  www.stevespanglerscience.com/content/experiment/how-to-make-a-folding-egg

- Teachers' Domain: Cell Membrane: Just Passing Through (lesson and simulation): *www.teachersdomain.org/resource/tdc02.sci.life.cell.membraneweb*

- Wikipedia: Osmosis: *http://en.wikipedia.org/wiki/Osmosis* (includes a visual model)

- Wolfram Demonstrations Project: Surface area increase by size reduction (animated): *http://demonstrations.wolfram.com/SurfaceAreaIncreaseBySizeReduction*

Answers to Questions in Procedure, steps #2–#5

2. *Engage:*

   a. Yes, all living things are made up of cells, with the possible exception of viruses, which are considered a somewhat unique category. Most cells are, in fact, microscopic. But an unfertilized egg is a single cell, although it lacks one-half of the required genetic material, so it cannot grow and develop into a living bird. Fertilized egg cells of any species are among the largest single cells, with most of the space occupied by food reserves for the initially microscopic but growing embryo. Human egg cells produced by females are quite large for human cells, about the size of a period—huge compared to male sperm cells.

   b. The cell membrane maintains an internal environment that is different from its external environment or "home neighborhood" by controlling inputs and outputs. Cell membranes operationally define "self" as distinct from "environment" at the molecular level. But the self is an ever-changing dynamic system undergoing constant repair and/or programmed death as determined by the cell's DNA.

   c. Yes, every egg has a cell membrane that is distinct from the shell. If desired, students can examine the membrane that remains attached to the shell of a cracked fresh and/or hard-boiled egg. Most students will not have previously noticed it or considered that eggs, like all cells (including those that have cell walls), have a cell membrane.

   d. Any acid will react with and dissolve the calcium carbonate shell.

3. *Explore:*

   a. Because the membrane appears flexible (like a balloon) and all membranes have molecular-level holes, it is likely that small molecules could pass through the membrane. An embryo would need to import oxygen and export carbon dioxide and water during cellular respiration that must occur 24/7 in every cell.

b. Water is the largest component of eggs, like all cells. Changes in water content could be noted by measuring the mass and/or circumference of the egg and the volume of liquid left in the container before and after the soaking. In the case of the de-shelled egg soaking in water, the water level in the container drops as the egg enlarges. Conversely, the corn syrup solution increases in volume slightly and becomes somewhat diluted. Typical initial egg circumferences may be 15–17 cm in the longer and 13–14 cm in the shorter dimensions. Subsequent changes in the 1–2 cm ranges are typical.

4. *Explain:*

   a. The meshing sorts "molecules" by size, allowing only the smaller ones to pass.
   b. The smaller objects that pass through the grid or "cell membrane" represent water, and the larger ones that cannot pass through represent larger biomolecules in the cell.
   c. In the model, gravity causes the particles to "fall down" (in one direction only) through the holes versus real cell membranes that are semipermeable in both directions.
   d. Water moves from a region of higher water concentration to one of lower water concentration; this concentration gradient is another type of energy gradient. Given the movement of other chemicals into and out of the cell, water has to be able to move in both directions to maintain homeostasis.

5. *Elaborate:*

   Part 2: The answers lie in the surface-area-to-volume scale effects that restrict movement of materials in and out of cells, which greatly restricts the size of prokaryotic cells. Eukaryotic cells increase their effective surface area by way of their many internal, membrane-bound organelles and rely on more complex active transport mechanisms more than "simpler" bacteria are able to do. Similarly, macroscopic, multicellular, eukaryotic organisms have innumerable ways of gaining surface area to volume and specialized structures and body organizational plans (e.g., villi in the small intestines and aveoli in the lungs). But surface-area-to-volume effects still influence body designs and physiological functions for organisms as small as shrews and as large as elephants and whales. The concept of scale is a big idea regardless of the size of the organism.

Activity 11

5 E(Z) Yet pHenomenal Steps to Demystifying Magic Color-Changing Markers

You may think I have a colorless personality but we could make pHenomenal chemistry together.

You may be blue now, but I can change that.

Expected Outcome

The chemical principles that explain the "science behind the magic" of color-changing markers are explored in a series of teacher-guided but learner-designed hands-on explorations.

Science Concepts

Color changes in materials can result from a simple physical mixing of pigments (e.g., food colors) or as a result of chemical changes associated with acid-base (pH), oxidation-reduction, photochromic, and thermochromic reactions. Many natural plant pigments are pH indicators or chemical substances that change color in response to changes in the relative concentrations of hydronium and hydroxyl ions. Chromatography is a physical separation technique that can be used to isolate the individual components of a mixture for subsequent identification. Macro-*scale* chemical and physical changes occur as the result of interactions of matter and energy in *systems* where certain factors stay *constant* (e.g., mass and the number of each type of atom and overall energy) and other factors *change* (e.g., relative molecular spacing and movement in physical changes versus interatomic bonding and molecular composition in chemical changes). Chemists explain these changes in terms of the atomic theory and molecular *models*.

Science Education Concepts

This activity is designed to model several principles of research-informed, best-practice teaching:

- The use of everyday phenomena or cultural artifacts (i.e., color-changing markers) as a starting point for instruction can raise questions that activate students' attention and catalyze cognitive processing.
- Laboratory instruction—though guided by the teacher with specific curriculum objectives in mind—can and should be more student directed, open ended, and exploratory than conventional follow-the-directions, "cookbook" exercises suggest.
- Learning that lasts is based on active and reflective two-way interactions between student and phenomena, student and student, and student and teacher that lead students to construct more refined and scientifically accurate mental models.
- The 5E Teaching Cycle can be a powerful means for sequencing curriculum-instruction-assessment (CIA) to achieve a desired end.

The chromatography simulation during the Engage phase can serve as a *visual participatory analogy* to show how different students

···5 E(Z) Yet pHenomenal Steps to Demystifying Magic Color-Changing Markers

progress at different rates depending on how their prior knowledge and abilities interact with the particular CIA system in place. Similarly, acid-base color-change chemical reactions can serve as an analogy for how the interactive nature of teaching-learning causes changes in both the teacher and learners and allows for the emergence of new ideas.

Materials

Crayola Color Switchers (*www.crayolastore.com/product_detail.asp? T1=CRA+58-8170&*; #58-8170; $5.99 for six, two-ended markers): One end of the marker appears to be a conventional color marker and the other end is colorless. When the colorless end is applied over the ink left behind by the colored end, a different color is formed. Crayola previously marketed a similar product under the name Color Changeables. Color Switchers are also available from arts and craft, drug, and department stores or as science lab kits (see p. 176). Various colors of standard water-soluble markers can also be tested to see how they react to the uncolored end of the Color Switchers marker. Each team of 2 learners will need 2 different-color markers.

Optimally, this 5E embedded laboratory experiment can be presented to learners as a somewhat open-ended exploration that uses primarily (or exclusively) household chemicals. Useful materials include cotton swabs (for dipping into liquids), red cabbage leaves for pH-sensitive rubs or juice (other indicators include colorless phenolphthalein indicator that turns pink in bases, litmus paper that turns from its red acid color to its blue base color, and universal pH paper that turns five different colors at different pH levels much like red cabbage leaves), and small volumes of various household acids (e.g., lemon juice [citric acid], vinegar [acetic acid/CH_3CH_2OH], and muriatic acid [hydrochloric acid/HCl, sold as a masonry cleaner in hardware stores) and bases (e.g., baking soda [sodium hydrogen carbonate/$NaHCO_3$], washing soda [sodium carbonate/Na_2CO_3], ammonia [NH_4OH], and lye [$NaOH$]) with different pH values. Chromatography can be done with chromatography paper, paper coffee filters, or paper towels cut into strips.

Optional: The 1939 film *The Wizard of Oz*, colored comic strips, and handheld microscopes. The microscopes are available from either of the following sources:

- *Horticulture Source*: EcoPlus Illuminated Microscope 30×, #704477 for $17.95 (uses 2 AA batteries): *www.horticulturesource. com/product_info.php/products_id/1194*

- *Radio Shack*: Illuminated 60–100× Microscope Model: MM-100, Catalog #63-1313; $12.49: *www.radioshack.com* (higher magnification/restricted field of view trade-off)

As an alternative to the more open-ended, student-designed exploration described below, several science suppliers have packaged kits with all the materials and laboratory worksheets:

- Carolina: Chroma-Tricks Experiment Class Chemistry Kit (#840447/$61.95): *www.carolina.com/product/chroma-tricks+experi ment+class+chemistry+kit.do*; 800-334-5551

- Educational Innovations Chroma-Tricks Chemistry Classroom Kit (#CK-465; $39.95): *www.teachersource.com/Chemistry/ChemistryKits/ Chroma_TricksChemistryExperimentKit.asp*; 888-912-7474

- Science Kit and Boreal Labs Chemistry of Color Change Kit (SKU WW65170M00;$77.00): *http://sciencekit.com/chemistry-of-color-change-kit-teacher-developed,-classroom-tested/p/IG0020015*; 800-828-7777

Points to Ponder

It is a cardinal precept of the newer school of education that the beginning of instruction shall be made with the experience learners already have … the educator views teaching and learning as a continuous process of reconstruction of experience … No experience is educative that does not tend both to knowledge of more facts and entertaining of more ideas and to a better, a more orderly arrangement of them … it goes without saying that the organized subject-matter of the adult and the specialist cannot provide the starting point. Nevertheless, it represents the goal towards which education should continuously move … It is a sound educational principle that students should be introduced to scientific subject-matter and be initiated into its facts and laws through acquaintance with everyday social applications … it is also the surest road to the understanding of the economic and industrial problems of present society.

—John Dewey, American philosopher-educator (1859–1952), in *Experience and Education* (1938)

····5 E(Z) Yet pHenomenal Steps to Demystifying Magic Color-Changing Markers

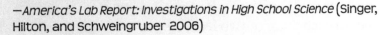

Points to Ponder *(continued)*

Most people in this country lack the basic understanding of science that they need to make informed decision about the many scientific issues affecting their lives. Neither this basic understanding— often referred to as scientific literacy—nor an appreciation for how science has shaped the society and culture is being cultivated during the high school years (p. 1) … Research focused on the goal of student mastery of subject matter indicates that typical laboratory experiences are no more or less effective than other forms of science instruction (such as reading, lectures, or discussion) … Due to a lack of available studies, the committee was unable to draw conclusions about the extent to which either typical laboratory experiences or integrated instructional units might advance the other goals identified … enhancing understanding of the complexity and ambiguity of empirical work, acquiring practical skills and developing teamwork skills (p. 5) … In general, most high school laboratory experiences do not follow the instructional design principles for effectiveness identified by the committee (p. 6).

—*America's Lab Report: Investigations in High School Science* (Singer, Hilton, and Schweingruber 2006)

Procedure

When Working With Teachers

A very abbreviated, less discovery-oriented version of the following 5E Teaching Cycle can be completed with teachers in as little as 30 minutes. The two Points to Ponder quotes can be shared as an introduction to help set a context for a critical assessment of the largely unrealized potential of laboratory-based learning, despite the 68-year gap between the two different quotes (from, respectively, an

education philosopher and the National Research Council of the National Academies).

When Working With Students

This 5E sequence can be completed as an open-ended series of hands-on explorations that extends over 2–5 class periods.

Engage:

1. Begin this multiday mini unit by asking the learners to scan their clothing and that of their peers and the objects around them in the classroom to see how many different colors they can spot. If time permits, share short contrasting video clips of an old black-and-white movie or TV show versus a modern, colored one to help students appreciate the "magic" that is color. Even better, you can use the famous twister segment from the 1939 film *The Wizard of Oz,* which begins in black and white and then dramatically shifts to color when Dorothy arrives in the land of Oz. Mention that the science of chemistry enables humans to explain, intentionally alter, and preserve the colors of materials, and explain that over the next several days students are going to systematically "play" with a commercially available arts and craft product, comic strips, and various household chemicals to see what they can discover about the chemistry of color.

2. Distribute two Crayola Color Switchers markers and plain white paper to each team of two learners and ask them to play with the markers to determine the range of colors that can be produced by using the colorless marker ends in combination with the various colored marker ends. If desired, the learners can use the markers to artistically complete the first two steps of a know—want to know—learned—plus (KWL; also includes graphic organizer such as a concept map) chart about the science of color. Ask the dyads to be systematic and record their tests, noting before and after colors. *Note:* Students should quickly discover that the second color on the caps "gives away" the color that forms when liquids from the two ends are mixed together and that all the colorless ends appear to contain the same liquid.

3. After allowing several minutes for the dyads to play with the two-ended markers, ask each dyad to pair with another team that has a different set of colored markers to share their results and brainstorm instances in nature, their homes, and chemistry labs where materials are observed to change color. Assuming they have already learned the difference, ask students to classify the Color Switchers and other examples as either physical or chemical changes (or a third category, "I don't know"). Share the results of the group brainstorming with the entire class and quickly review the definitions of physical versus chemical changes (without giving students answers to whether the observed change was physical or chemical) and their KWL charts.

4. *Optional Mini-Exploration: The Colors of Comics:* Distribute handheld microscopes and sections of both the daily black-and-white and Sunday color comic strips to each dyad. If students have not already used the handheld microscopes, quickly show them how to do so and let them discover the apparent reversal of up and down and left and right. Then, have them compare and contrast naked-eye versus magnified-image observations of the comic strips. Ask them to pay particular attention to the composition of the different individual colors and the black ink in the black-and-white strips versus that found in the colored comic strips (i.e., discover what is "funny about the funnies"). It is also interesting to examine the faces of the models featured in colored magazine ads for skin products with the handheld microscopes. *Note:* Although light microscopes cannot reveal individual molecules, the idea that there is a level of organization beneath the visual level of the unaided human eyes that can be quite different opens students' minds to the possible realm of atoms and molecules.

Explore:

Focus Question: How can the colorless "ink" cause a color change in a colored marker?

Drawing on students' brainstorming and categorized list, teams of four should develop at least two different hypotheses and a plan to test each as possible explanations for the observed color changes with the markers. Depending on the students' prior knowledge about

acid-base chemistry and their experience with more open-ended labs, they will need different levels of teacher scaffolding to support their inquiry. Also, after pooling the ideas of all the groups in a whole-class discussion, the teacher may wish to assign different teams to do different experiments (to save time). In this case, be sure to assign at least two teams for each test performed as a reliability check and to support collegial skeptical review. Consider the following suggestions as possible hints to be used as needed.

1. There is something "special" about the pigments in the *colored* ends of the markers. Often, what appears to the naked eye to be a single color is in fact a mixture of different color components (i.e., analogous to the comic strips). Perhaps the *colored* ends of the markers contain multiple dyes that are mixed together (like the different color pigments in leaves, food coloring, and regular water-soluble colored markers) that can only "show up" under specific "environmental" conditions. *Test:* Paper chromatography can be used to try to separate possible "hidden" colors in the Color Switchers and other "regular" water-soluble markers (i.e., it is especially interesting to test various brands of black markers). Teams should use their textbook and/or the internet (a few sample sites are included below) to find out how to run chromatograms. Also, if separation occurs with any of the colors, the students can test the effect of the colorless marker on the separated components to see which components change color in the presence of the colorless liquid on the other end of the markers.

2. There is something "special" about the *colorless* end of the markers. Perhaps the *colorless* ends of the markers contain an acid or base that chemically reacts with and changes the pH of the pigments in the colored ends of the markers (that contain one or more "indicator dyes").

   a. Check the effect of various household acids and bases (see examples under Materials) on the colored markers to see if one or more of these can replicate the effect of the colorless end of the markers on one, more, or all of the colored markers.

   b. Use phenolphthalein, litmus paper, and/or universal pH paper (or better, red cabbage leaf "rub" as a great five-color, natural universal indicator) to test the pH of the liquid in the colorless end of the markers.

c. Depending on the measured pH of the colorless ends of the markers, check if either a household "anti-base" (i.e., acid) or "anti-acid" (antacid or base) can reverse the colorless marker's effect and return the colored marker to its original color.

d. Test if the colorless marker will effect a color change when used with regular water-soluble markers that are not sold as Color Switchers.

e. Check to see if the order that the markers are painted on the paper matters.

3. The something "special" is the result of chemical interactions and reactions between the different chemicals in the two ends of the marker to produce new chemical compounds with different properties. One would need to complete all the previously listed tests to support this correct, but general, hypothesis.

Explain:

After the teams have had at least one class period to test their ideas in the laboratory, pool and discuss their empirical evidence and challenge the teams to use logical arguments and skeptical review to develop a tentative, first-level explanation for the science behind the magic markers. All macroscopic physical and chemical changes in matter have corresponding explanations at the level of molecules, their movements and atomic compositions, and specific 3D configurations. Although science teachers have long ago internalized the idea of molecules, for most students they remain a somewhat abstract, invisible magical rabbit that teachers pull out of the hat to explain macroscopic changes. Getting students to visualize molecules and their physical movement is an essential first step to getting them to understand the breaking and making of chemical bonds and the associated changes in energy that together account for all chemical changes. Though other activities in previous books in this series (*More Brain-Powered Science*, activities #8, #15, and #17; *Brain-Powered Science*, activities #14, #16, #20, and #27) present empirical evidence, logical arguments, and skeptical review to "build the case" for the kinetic molecular theory (KMT), the following interactive

participatory demonstration asks students to role-play molecules-in-motion.

Modeling Molecular Movement Matters: As a relatively simple physical change, paper chromatography is a great entry point into the unseen realm of molecules. One aspect of the unseen underlying chemistry of chromatography can be modeled as follows. Before class, put a set of different color but otherwise identical objects in an opaque black bag (e.g., sponges are commonly sold in dollar stores as packs with a rainbow of colors represented). During class, inform students that this bag represents the chemical mixture of a water-soluble black ink marker and ask: What might be in this black bag to represent the results you found during your chromatography experiments? If we use these objects as a macroscopic, visual participatory analogy for the different chemical compounds in the ink mixture, what kinds of rules are needed to mimic the molecular "race"?

Open the bag, pull out the colored sponges (or other objects), and give one colored sponge to each person in a group of four to six students. Collectively set the "unfair" rules for their respective linear movement to match how the colors separated from the black ink. For example, the first student/sponge can move 0.5 m in a given time interval (or a certain number of tile blocks in the floor), the second moves 1 m, the third 1.5 m, the fourth 2 m, the fifth 2.5 m, and the sixth 3 m. *Note:* These rules model the different degrees of attraction (i.e., adhesion and cohesion) that the different chemical components have toward the water solvent and the chromatography paper that leads to their separation during the macroscopic version of the molecular "race." Ask for a student volunteer to provide a constant cadence of sound by regularly clapping hands or snapping fingers at a rate that his or her colleagues could move from a common starting line where they are initially "mixed together." This may involve a little playful exploration, but the students will easily see that the mixture that was initially black will separate into different colors just like the black ink on the chromatograph paper. For a more artistic variation of chromatography, see Internet Connections: Steve Spangler Science: Sharpie Pen Science for a description of how to make tie-dyed T-shirts.

Of course, the most dramatic aspect of the color change markers is the acid-base chemical reaction between the colored pigments (indicators) and the other colorless, alkaline (basic) end of

···5 E(Z) Yet pHenomenal Steps to Demystifying Magic Color-Changing Markers

the markers. The following curriculum-embedded, formative assessment can be used during the Explain phase to help gauge how much the students already know about the underlying chemistry involved in these experiments so that the teacher can determine how to best frame explanations in subsequent lessons.

ABCs of Acid-Base Chemistry, Color Changes, and Chromatography

The following nongraded "test" is designed as an indicator of the science you already know about the phenomena that we have been exploring. The results will help us assess what knowledge gaps and/or misconceptions we need to address.

		True	False	Unsure
1.	A good way to test an unknown substance to determine if it is an acid is to taste it to see if it is sour.	T	**F**	U
2.	All acids are harmful chemicals that "eat" metals and burn the skin.	T	**F**	U
3.	Bases are a class of chemicals that work as "anti-acids."	**T**	F	U
4.	Many plant pigments will change color in the presence of a sufficiently strong acid or base.	**T**	F	U
5.	Chemicals are always liquids or gases that are manufactured in labs.	T	**F**	U
6.	A pH indicator is a chemical that changes color in the presence of acids or bases.	**T**	F	U
7.	Physical changes result in the formation of new substances with different physical and chemical properties.	T	**F**	U
8.	Rainwater, even in unpolluted air, is naturally acidic.	**T**	F	U
9.	A good way to test an unknown substance to determine if it is a base is to see if it has a soapy feel when rubbed between your fingers.	T	**F**	U
10.	Acid strength is the same as acid concentration.	T	**F**	U
11.	The ingredients used in cleansing agents, cosmetics, and medicines are often acids or bases.	**T**	F	U
12.	Most foods are chemically neutral—neither acidic nor basic—and therefore are safe to eat.	T	**F**	U
13.	Solids dissolved in water form solutions that are acidic, neutral, or basic as measured on the pH scale.	**T**	F	U
14.	Pure water is neutral; it is neither acidic nor basic.	**T**	F	U
15.	pH is a linear scale, so a pH of 2 is twice as acidic as a pH of 1.	T	**F**	U

16.	Some gases can dissolve in and react with water to form acids.	**T**	F	U
17.	All macroscopic (visible) chemical changes in matter are explained by the breaking of old and making of new chemical bonds between different atoms at the (sub-) microscopic scale.	**T**	F	U
18.	Chromatography causes chemical changes in a mixture.	T	**F**	U
19.	A mixture can be separated by physical means.	**T**	F	U
20.	All color changes involve chemical reactions.	T	**F**	U
21.	Chemicals are involved any time matter changes color.	**T**	F	U
22.	A question that I have about acids and bases is ...			
23.	A question that I have about the chemistry of color is ...			

Whether or not this assessment is used, provide students access to grade level–appropriate physical science and chemistry textbooks and the internet so they can track down more information on paper chromatography, acid-base indicators, and perhaps even the specific chemistry behind the color-change markers. When it comes to more teacher-directed presentations of the underlying science, for middle school students it is appropriate to leave the explanation for the change at the level of the color changes as a result of acid-base chemical reactions (i.e., the pigments are one color in their more acidic state and change color when the pH changes to basic when the colorless marker is used). At the high school chemistry level, the PhET Interactive simulation (Internet Connections) can be used to "test" the pH of 11 different household substances (including bodily fluids) and see graphic displays of the relative concentrations of hydronium and hydroxyl ions. If desired, the role of oxidation-reduction chemistry (chemical "bleaching" such as that caused by household bleach [sodium hypochlorite/NaOCl]) can be explored by challenging the students to "dissect" the text of the actual original patent behind this product (see Internet Connections).

At either grade level, it is important to use the students' results and thoughts as the basis for a more refined, teacher-directed explanation targeted at the appropriate conceptual level. Be sure to emphasize everyday household foods and other chemical products that are acids and bases and how the pH contributes to their taste (for foods) or effectiveness in specific applications; the importance of pH in biochemistry and health; and the chemical safety precautions that

• • • 5 E(Z) Yet pHenomenal Steps to Demystifying Magic Color-Changing Markers

consumers should follow when using household chemicals. It is also important that students learn that the pH scale is not a linear scale (but based on powers of ten or a logarithmic scale) and that all acid (or bases) are not the same (e.g., vinegar is a much weaker acid than hydrochloric acid, and sodium hydrogen carbonate is less basic than ammonia, which is in turn less basic than sodium hydroxide). In addition to litmus and red cabbage, a wide range of colored plant pigments work as acid-base indicators. *Note:* All four of the *Benchmarks for Science Literacy* (AAAS 1993) common themes of systems, models, constancy and change, and scale can be explicitly emphasized in the discussions related to acid-base chemistry (see Appendix C).

Elaborate:

One or more of the following lines of investigation can be pursued as desired by either the class as a whole or student-selected self-interest groups using laboratory- and/or internet-based investigations.

1. *STS Investigation: It's Raining Acid!* Since the colorless marker was shown to be a base, is there any component of air that could potentially reverse any of the observed color changes over time? (*Note:* The colorless end of the marker changes universal pH paper to blue, indicating a pH of 10 or more. As it dries and is exposed to carbon dioxide in the air, the pH paper's color turns to green [pH = ~ 8]). Is clean, unpolluted air a neutral solution? Why or why not? What is "acid rain," what are the direct and indirect human causes of acid rain, and why should acid rain be a cause for concern for all citizens (not just environmentalists)? Students can use a variety of indicators that change color in the pH range of 6–8.5 (e.g., bromthymol blue and phenol red) to demonstrate that the carbon dioxide produced in respiration forms a weakly acidic solution. Depending on the indicator used, the water may first need to be made slightly basic. This reaction is not caused by "bad breath," but the color change will happen faster if the person does vigorous aerobic exercise before exhaling through a straw into the indicator solution (with the safety caution to not inhale; see Internet Connections: Biology Corner). The same color-change reaction can be observed by using dry ice (solid carbon dioxide) or the gas given off by a candle. Of course, the

main sources of "unnatural" atmospheric acidity are nitrogen oxides (from high-temperature combustion, which causes a reaction between the nitrogen and oxygen in air) and sulfur oxides (from sulfur-containing fossil fuels). If a fume hood is available, teachers can demonstrate that these gases form much stronger acidic solutions than carbon dioxide does.

2. *pHenomenal Plant Pigments, Paper, and Foods:* Explore a variety of colored plant pigments (e.g., red cabbage, grape juice, beet juice, turmeric spice, etc.), goldenrod colored paper (specially purchased from chemical supply companies such as Educational Innovations and Steve Spangler Science), and foods for their potential use as acid-base indicators, their tastes, and the roles they play in recipes due in part to their pH values. See Activity #2, the Internet Connections below, and *More Brain-Powered Science*, Activity #12.

3. *pHenomenal Pools and Purification of Water:* Explore other commercially available acid-base indicators used for water testing that are sold in pool and pet and aquarium stores and why the pH of water is so critical to the health and well-being of humans, other animals, and plants. Virtual or real field trips to water purification plants can bring the relevance of pH to center stage for students.

4. *Consumer Product Testing: Crayola Color Wonder Paper (www.crayola. com/products/splash/color_wonder/index.cfm?n_id=56):* This product is sold as a set of six clear, colorless markers that show up as colored ink *only* when they are used on Crayola's Color Wonder paper. Students can investigate how this product works.

Evaluate:

Though conventional paper-and-pencil assessments can be used as part of this chemistry mini- unit, the possibilities for authentic, performance-based assessments are numerous. In addition to using any of the ideas listed under the Elaboration phase, consider ideas such as these:

1. Have students design a "Chemistry Is Magical But Not Magic" demonstration, a hands-on program for younger students in your

school district, or a presentation for the PTA. Both acid-base chemistry and chromatography lend themselves to colorful presentations that can be easily targeted for different age groups.

2. Students can create PowerPoint presentations for their classmates that explain other real-world examples and applications of either chromatography (which could be expanded to include gas and high-pressure liquid chromatography) or acid-base chemistry. Both of these assessment alternatives are based on the idea that if you really want to learn something, try to teach it to someone else.

Debriefing

When Working With Teachers

Consider how physical and chemical changes can serve as visual participatory analogies for the interactive and change-oriented nature of teaching-learning in schools that are learning organizations for both teachers and students. Return to the two Points to Ponder quotes as discussion prompts for a critical examination of the current state of laboratory exercises (versus explorations, investigations, or invitations to inquiry). Interested teachers may wish to read the executive summary of NRC's *America's Lab Report* (see Singer, Hilton, and Schweingruber 2006 or the Internet Connections below), which provides the following simple definition: "Laboratory experiences provide opportunities for students to interact directly with the material world (or with data drawn from the material world) using the tools, data collection techniques, models, and theories of science" (Singer, Hilton, and Schweingruber 2006, p. 3). As such, NSTA (see Internet Connections) and other science organizations have always been staunch supporters and active promoters of the integral role of laboratory-based learning. However, the National Research Council's report echoes Dewey's critique by noting that most lab-based learning fails to meet one or more of the following four key principles of instructional design:

- Have clear learning outcomes
- Are thoughtfully sequenced into the flow of classroom science instruction (i.e., are part of *integrated instructional units* such as a 5E Teaching Cycle)

- Integrate the learning of science content and processes
- Incorporate ongoing student reflection and discussion.

As such, this report concludes, "The quality of current laboratory experiences is poor for most students" (Singer, Hilton, and Schweingruber 2006, p. 197).

Commercial textbook publishers are starting to include better-designed lab activities, so teachers can and should critically examine and modify conventional "cookbook labs" in terms of how they represent the nature of science (NOS); reflect an inquiry orientation; provide real-world relevance and motivation; connect to their state's grade-level science curriculum and/or national standards; and are appropriate in terms of students' developmental levels, safety and waste disposal issues, expense, and their overall cost- or risk-to-benefit ratio. Practical support for teachers interested in improving lab activities is available in a number of publications (Clark, Clough, and Berg 2000; Clough 2002; Clough and Clark 1994a, 1994b; Colburn 2004; Doran et al. 2002; Hofstein and Lunetta 2003; Lunetta, Hofstein, and Clough 2007; NSTA 2007c; Volkman and Abell 2003). In the age of word processing, e-mail, and the internet, laboratory learning is a wonderful topic for teachers-helping-teachers professional development, where progressive cycles of shared action research can be used to polish "rough-cut gems into highly polished diamonds." Keep in mind that something as simple as where a teacher chooses to place a given lab activity (i.e., preferably in an Explore phase before rather than after the Explain phase that gives students the "right answers"); the wording of the purpose statement; and the connections to and applications in real-world, out-of-school contexts can make a given activity more or less inquiry-based, motivational, and pedagogically valuable.

When Working With Students

A formal laboratory write-up was intentionally omitted because the intent of the activity is to challenge the teacher-readers to consider running this hands-on exploration in a more open-ended, discovery-oriented way. Depending on the grade level and prior knowledge of the students, teachers may elect to provide more scaffolding and guidance to support their students' investigations. The formative assessment that is built into the Explain phase can be analyzed during the Elaborate or Evaluate phases as a means to model for students how to critically analyze true-false items (e.g., for a statement to be true, all of the statement must be true; one word can make a difference, so careful reading is required to minimize errors). Although correct answers are provided for this test, the author elected not to include any Answers to Embedded Questions, but rather only key references in the Internet Connections. The intent here is that the teacher-reader will have the opportunity to investigate this phenomenon on their own without being burdened or disadvantaged by knowing the answers ahead of time.

Extensions

See the Elaboration phase on page 185, Activity #2, the Internet Connections (below), and *More Brain-Powered Science*, Activity #12, "Magical Signs of Science," for other activities that involve acid-base indicators and/or chromatography.

Internet Connections

• About Chemistry:

> Acid-Base Indicators, their pH ranges and respective colors: *http://chemistry.about.com/library/weekly/aa112201a.htm*

> Chemistry Experiments and Demonstrations You Can Do at Home (e.g., make disappearing ink): *http://chemistry.about.com/od/homeexperiments*

• *America's Lab Report: Investigations in High School Science*. Executive Summary: *www.nap.edu/catalog/11311.html*

• Biology Corner: Scientific Method: Carbon Dioxide Production (lab): *www.biologycorner.com/lesson-plans/scientific-method*

• Chem Guide: Indicators: *www.chemguide.co.uk/physical/acidbaseeqia/indicators.html*

> Paper Chromatography: *www.chemguide.co.uk/analysis/chromatography/paper.html*

• Chemistry in the Library: What's in a Color (lab): *www.mdchem.org/citl/CitL-Color_Expt.pdf*

• *ChemMatters* CD (1983–2008): *http://portal.acs.org/portal/acs/corg/content*; 800-227-5558

Search for CMCD4 ($30.00); full searchable, printable, and includes Teacher's Guide. This ACS-produced magazine is written for high school chemistry students and contains a wealth of real-world relevant chemistry. The October 1998 issue [16 (3)] contains an article (pp. 4–6) by Cynthia Anderson and David Katz titled "A Mark of Color" that explains the chemistry behind the magic of the color-changing markers.

- David Katz: Investigating the Chemistry of Color Changing Markers: *www.chymist.com/color%20changing%20investigation.pdf*

- *Democracy and Education*: 1916 book by John Dewey: Columbia University's Institute for Learning Technologies' Digital Text Project: *www.ilt.columbia.edu/Publications/dewey.html*

- Disney Educational Productions: *Bill Nye the Science Guy*: *Chemical Reactions* ($29.99/26 min. DVD): *http://dep.disney.go.com*

- Doing Chemistry: Movies of chemical demonstrations: Colorful reactions of red cabbage juice:

  http://chemmovies.unl.edu/chemistry/dochem/DoChem098.html

  http://chemmovies.unl.edu/chemistry/dochem/DoChem099.html

- Fun Science Gallery: Experiments With Acids and Bases (red cabbage): *www.funsci.com/fun3_en/acids/acids.htm*

- General Chemistry Online: Indicators and Their Color Changes/pH Ranges: *http://antoine.frostburg.edu/chem/senese/101/acidbase/indicators.shtml*

- Goldenrod Paper pH Indicator:

  Becker Demonstrations: *http://chemmovies.unl.edu/chemistry/beckerdemos/BD022.html*

  Daryl's Science: *www.darylscience.com/Demos/Goldenrod.html*

  Educational Innovations: *http://blog.teachersource.com/2009/10/01/goldenrod-paper*

  Science Hobbyist: *http://amasci.com/amateur/gold.html*

  Steve Spangler Science: *www.stevespanglerscience.com/experiment/00000040*

····5 E(Z) Yet pHenomenal Steps to Demystifying Magic Color-Changing Markers

- National Science Center: How Color Change Markers Work: Video: *www.nscdiscovery.org/TeacherTools/topic_details.asp?ID=184*

- National Science Teachers Association Position Paper: *The integral role of laboratory investigations in science instruction*: *www.nsta.org/about/positions/laboratory.aspx*

- PhET Interactive Simulations: *http://phet.colorado.edu/simulations/sims.php?sim=pH_Scale*

- Purdue University: Voice-activated chemical reaction: *www.chem.purdue.edu/bcce/Voice_Activated_Reaction.pdf*

- Steve Spangler Science:

    Sharpie Pen Science (tie-dyed t-shirts): *www.stevespanglerscience.com/experiment/00000032*

    Do You Have Acid Breath?: *www.stevespangler.com/experiments*

- University of Virginia Physics Dept: pH, Soda Science, Flower Indicators, and Magic (HOEs):

    http://galileo.phys.virginia.edu/outreach/8thGradeSOL/SodaScienceFrm.htm

    http://galileo.phys.virginia.edu/outreach/8thGradeSOL/FlowerIndicatorsFrm.htm

    http://galileo.phys.virginia.edu/outreach/8thGradeSOL/AcidBaseFrm.htm

- U.S. Patents Office: Patent #5,232,494: Color changing compositions by Richard E. Miller of Binney & Smith Inc.: *http://patft.uspto.gov/netahtml/PTO/srchnum.htm*

- Wikipedia: *http://en.wikipedia.org/wiki*. Search topics: autumn leaf color, chromatography, pH indicators (includes list), and redox indicator.

- *YesMag: The Science Magazine for Adventurous Minds*: Paper Chromatography: *www.yesmag.ca/projects/paper_chroma.html*

5 E(Z) Steps Back Into "Deep" Time

Visualizing the Geobiological Timescale

A ROLL OF TOILET PAPER as an EVOLUTIONARY TIMELINE?

Expected Outcome

A sequence of fun, participatory activities juxtapose every-day popular culture and human time perspectives with the geobiological timescale of Earth's history of millions and billions of years. The overall timescale of Earth's history as a planet, the sequence of events in its ongoing evolution-driven story of life, and the minis-cule percentage of Earth's history represented by written human history are all quite discrepant to most learners.

Note: This activity has been modified from the original source:
O'Brien, T. 2000. A toilet paper timeline of evolution: A 5E cycle on the concept of scale. *The American Biology Teacher* 62 (8): 578–582. Permission from the National Association of Biology Teachers (NABT).

Science Concepts

Earth's interwoven longer-running geological and more recent bio-logical evolutionary histories are both scientific facts and a set of interrelated scientific theories that have powerful explanatory and predictive power. The ideas of *constancy* and unity (i.e., order or "cosmos") amidst *change* and diversity (i.e., seeming disorder or "chaos") have evolutionary, conceptual roots that date back to at least the time of the ancient Greek natural philosophers. But the literal and figurative groundbreaking work of biologists, geologists, and paleontologists in the late 1700s to early 1800s laid some of the cornerstones of the empirical foundation for Darwin's development of the theory of biological evolution (or descent with modification) by natural selection. Understanding that Earth is "very old" (i.e., orders of magnitude older than some biblical literal interpretations of approximately 6,000 years) and the human species is "very young" is a conceptual prerequisite for both geological processes (e.g., plate tectonics, mountain building, earthquakes, and volcanoes) and bio-logical evolution to "make sense."

Being able to visualize and conceptualize the powers of ten *scale* differences in the magnitudes of individual human life ($\sim 10^2$), writ-ten records of human history (10^3–10^4), *Homo sapien* history ($\sim 10^5$), history of life (10^6–10^9), and the age of the Earth (10^9)—as devel-oped in this mini unit—builds the foundation for subsequent units such as the following:

1. Earth science: plate tectonics, half-life of naturally occurring radioisotopes, sedimentation, and fossil formation

2. Life science: biological classification, comparative anatomy and physiology, and biostratigraphy that allows paleontologists to "read the story" of the relative time-ordered sequences of fossils that link extant life forms to distant ancestors (most of whom are now extinct). Any units related to evolution also provide oppor-tunities to focus on the nature of science (NOS) concepts, such as its intentionally agnostic methods and explanatory models and to clarify misconceptions about the role of empirical evi-dence, facts, proof, and theory.

Science Education Concepts

Evolution is both a unit-level topic and a course-long theme that should play the same central role in middle and high school Earth and life sciences curricula as it does in these sciences themselves (AAAS 1993; NRC 1996; NRC 2010a—see Appendix C). No single, several-day-long 5E Teaching Cycle (such as this one) can do this central organizing theory justice with respect to the breadth of inter-related strands of supporting empirical evidence, the strength of its logical arguments, and its ability to withstand the test of more than 150 years of skeptical review and ongoing research. The theory of evolution of the Earth system, including both its geological and bio-chemical subsystems, is challenging for both teachers and students due to a number of synergistically linked conceptual, historical, methodological, pedagogical, philosophical, political, and theological factors (Miller 1998; National Academy of Sciences 1998, 1999, 2008—see Resource Books and Articles for Teaching Evolution on pages 215–216).

The following sequence of activities focuses on one critical part of the pedagogical problem presented by evolution, the scale of "deep time," via the use of physical, mathematical, and analogical models. Factors that make it difficult to teach or understand the idea of deep time include those listed:

- Geobiological time charts divide Earth's continuous history into variable-length time segment blocks that are dated backward from the present (as millions of years ago [mya]) rather than forward (like birthdays) from the origin of Earth, or "time zero."
- Foreign-sounding geological terminology for the various eras, periods, and epochs of Earth history is compounded with off-scale, truncated tables or charts with broken timelines.
- The scale of millions and billions lies outside most learners' experiences.
- The seemingly rock-solid, immovable ground of Earth's litho-sphere and the composition of the atmosphere and hydrosphere have all changed dramatically across this immense timescale, cre-ating very different environmental conditions for species to adapt (or fail to do so) at different times.

• Evolution within a given species and the biogenesis of most "new" macroscopic species is a very slow process from a human time perspective.

Other published activities (see Resource Books and Articles for Teaching Evolution on pp. 215–216) make use of either relatively long scales (a football field [McComas 1994]) or scales that are rather short (5 m of adding tape [National Academy of Sciences 1998, pp. 90–92]); this activity uses an intermediate length that requires getting out of the classroom but not outside the school building. The following 5E Teaching Cycle develops the timescale of evolution in a way that goes beyond linguistic and logical-mathematical limitations by including musical, spatial, bodily kinesthetic, and personal intelligences (Armstrong 2000). It also features all four of the Benchmarks common themes: systems, models, constancy and change (e.g., evolution), and scale (AAAS 1993; see also Appendix C).

Life evolves via "creative" modifications of pre-existing variations (i.e., mutations and sexual recombination of genes), competition between these variations, "survival of the fittest," and extinction of species that are "unfit" for particular but changing environments. In an analogous fashion, learners' mental schema (i.e., conceptual networks of interrelated ideas) can be viewed as reproducing, competing for survival (with other ideas), and generating "offspring" that are either better adapted to changing conditions or eliminated due to their relative lack of "fitness" (i.e., their ability to both explain past phenomena and predict new ones). Simply put, our mental schema can be viewed as constantly evolving under selection pressure from our external learning environment. The task for teachers is to design and refine (during classroom implementation) an intelligent, adaptive curriculum-instruction-assessment (CIA) system that supports learners' next-step conceptual (re-)construction by building on and extending their prior experiences and valid conceptual precursors; modifying or forcing the extinction of unfit misconceptions; and giving birth to new, improved species of thought that better fit the environment of current scientific understanding. Individual learning is a process of conceptual change that—like the "evolution" of scientific theories— takes time, requires the proper environmental selection pressures, generally involves some "miss-takes," and is never finished evolving. Learners cannot leap several hundred or thousand years of scientific thought in a single bound without cognitive scaffolding to help them evolve from more primitive prior ideas.

Materials

- Recorded music (or access to YouTube videos) of a sample of popular songs from the 1950s through present that feature time in the title and lyrics (see Engage phase below for possible songs)
- Science cartoons that feature geological and biological evolution (see Internet Connections)
- A class set of 5 in. × 8 in. Earth Events cards (see table at end of this activity); one event per card is needed for each learner in the class.
- 1 new roll of toilet paper per class

Procedure (For Teachers and High School Students)

Share one or more of the Points to Ponder quotes to help set a context for this multilesson instructional sequence. Depending on the setting (i.e., grades 5–12 classrooms or teacher professional development context), one can spend as much as a full class period (or more) for each teaching phase or rapidly model selected highlights of the entire 5E sequence (with teachers who already know the content) in as little as 30 minutes.

1. *Engage:* Use one or more of the following humorous strategies to activate learners' attention and motivate them to think about both absolute and relative time from human versus biological and geological timescale perspectives:

   a. *Guess-My-Age Game:* Invite learners to guess the implied age of randomly presented chronologically challenged or generationally gifted birthday T-shirt messages such as these:

      - "Clear the road, I'm __ ." and "Baby on board and he's driving the car." (16)
      - "Go ahead and card me, I'm ___ or 7,665 days (or 183,960 hours or 11,037,600 minutes) old, but who's counting?" (21)

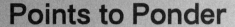

Points to Ponder

Nothing in biology makes sense except in the light of evolution.

—Theodosius Dobzhansky, Ukranian American geneticist and evolutionary biologist (1900–1975), from an article in *The American Biology Teacher* 35 (1973): 125–129.

In the twentieth century, no scientific theory has been more difficult for people to accept than biological evolution by natural selection. It goes against some people's strongly held beliefs about when and how the world and the living things in it were created. It hints that human beings had lesser creatures as ancestors, and it flies in the face of what people can plainly see—namely that generation after generation, life forms don't change (p. 122) … Most variables in nature … show immense differences in magnitude … A million becomes meaningful, however, as a thousand thousands, once a thousand becomes comprehensible. Particularly important senses of scale to develop for science literacy are … the enormous age of the Earth and the life on it. (p. 276)

—*Benchmarks for Science Literacy* (AAAS 1993)

In studying the evolution of the Earth system over geologic time … [to] … unravel the interconnected story of Earth's dynamic crust, fluctuating climate, and evolving life forms … The challenge of helping learners to learn the content of this standard will be to present understandable evidence from sources that range over immense time scales … Many students are capable of doing this kind of thinking, but as many as half will need concrete examples and considerable help in following the multistep logic necessary to develop the understandings.

—*National Science Education Standards* (NRC 1996, p. 188)

Points to Ponder *(continued)*

Earth is a complex and dynamic 4.6 billion year old system of rock, water, air and life … and life began almost 4 billion years ago. Thus, both Earth and life have developed over a HUGE expanse of time. Millions and billions of years are nearly impossible to comprehend from a human time perspective (p. 7–29) … Biological evolution explains both the unity and diversity of species … The fossil record documents the existence, diversity, extinction and change over time of many life forms throughout Earth's history … [and] led to the idea that newer life forms descended from older life forms (p. 7–18).

—A Framework for Science Education: Preliminary Public Draft (NRC 2010)

In an ideal world, every science course would include repeated reminders that each theory presented to explain our observations of the universe carries this qualification: "as far as we know now, from examining the evidence available to us today."

—Murray Gell-Mann (b. 1929–), American physicist ("quarks and strangeness")

- "I will be 29 until further notice and I'm still a teenager in Galapagos tortoise years." (30)
- "I'm 280 years old in dog years and I reached 20 a second time!" (40)
- "I'm not __, I'm 18 with 32 years of experience." "Half a century young." (50)
- "Just a baby in bristlecone pine tree years" (60)

Discuss how *old* is a relative term that depends on one's perspective as an individual and a species (e.g., dogs live anywhere from 7–16 years, typically inversely related to the size of the breed; Galapagos tortoises live as long as 150 years; and bristlecone pines can live nearly 5,000 years). Different species also reach reproductive maturity at very different rates.

b. *Sequence-the-Songs Game:* Play a randomly mixed compilation of song clips (and/or YouTube video clips) from recent decades that feature the idea of time:

- "Rock Around the Clock" (Bill Haley and His Comets 1954)
- "Any Time at All" (Beatles 1964)
- "The Times They Are a-Changin'" (Bob Dylan 1964)
- "Turn, Turn, Turn!" (Byrds 1965)
- "Time of the Season" (Zombies 1967)
- "Long Time Gone" (Crosby, Stills, and Nash 1969)
- "Good Times Bad Times" (Led Zeppelin 1969)
- "In the Year 2525/Exordium and Terminus" (Zager and Evans 1969; projects future human evolution—see YouTube slide shows)
- "No Time" (The Guess Who 1969)
- "Does Anybody Really Know What Time It Is" (Chicago 1969)
- "Time in a Bottle" (Jim Croce 1972)
- "Time" (Pink Floyd 1973)
- "Past Time Paradise" (Stevie Wonder 1976)
- "Long Time" (Boston 1977)
- "Longer Than" (Dan Fogelberg 1980)
- "The Best of Times" (Styx 1981)
- "It's About Time" (John Denver 1983; features an ecological theme)
- "Caught Somewhere in Time" (Iron Maiden 1986)
- "Time Passes By" and "Lonesome Standard Time" (Kathy Mattea 1991 and 1992)
- "Time Precious Time" (Lindsey Buckingham 2008)

Ask learners to put the songs in the correct time order and guess the year (or decade) of the release. If desired, analyze select lyrics that can be easily linked to the idea of evolution. Briefly discuss how a diversity of song styles may exist (and compete with one another for commercial "survival") at the same time and how older songs may exhibit the antecedents of modern elements—not unlike the unity (i.e., biochemical relatedness and common origin) and diversity (i.e., 10–100×10^6 living species) of the evolutionary story of life itself.

c.   *What's So Funny About Science Cartoons?* Display or distribute representative evolution-related science cartoons by Sidney Harris, Gary Larson, and others (see Internet Connections and related books by the same cartoonists). Invite learners to "dissect and discuss" how these cartoonists play with the nature of science and the concept of Earth's geological and evolutionary time (e.g., the intentional use of incorrectly paired, "contemporary" species such as dinosaurs and humans). Note that what is considered funny is relative to individuals' knowledge base and perspective, which can depend on age.

Use the following Focus Questions to help learners make the transition from everyday timescales of human years, decades, and lifetimes to "Earth time":

Focus Question 1: How can powers of ten (10^x) be used to represent the differences in magnitudes of individual human life, written records of human history, *Homo sapien* history, and history of life on Earth? How do scientists "read" the historical record of "deep time" that predates the human-written records studied by historians?

Focus Question 2: Is the age of the Earth measured in thousands (10^3), ten thousands (10^4), hundred thousands (10^5), millions (10^6), or billions (10^9) of years? What ways do scientists have of estimating the Earth's "birthday"? And so what? Who cares? Or why should I care?

Tally the learners' guesses about the age of Earth and note that one of the biggest challenges to Darwin's theory of biological evolution when he first proposed it was that scientific estimates at that time for Earth's age were "only" in the range of about 24–100 million years (see Internet Connections: Contextual Science Teaching). The lower estimate by the famous British physicist William Thomson (Lord Kelvin) was dramatically increased after the late 1890s and early 1900s discovery of radioactivity within the Earth's core as a source of ongoing internal heat. Provisionally offer the modern estimate of 4.55 billion years for Earth's age. Briefly discuss how scientific educated guesses are typically arrived at by a triangulation of data from multiple sources, analogous to intersecting words on a crossword puzzle or how multiple satellites are used by GPS devices to pinpoint location. Science relies on the triad foundation of converging empirical evidence, logical argument, and skeptical review. Without going into details (reserve these for the Explain phase if time allows), note that fundamental concepts and theories in astronomy, biology, chemistry, climatology, geology, paleontology, and physics have all contributed to and are dependent on the idea of a "very old and ever-evolving" Earth system as indicated by radiometric dating and other techniques.

Focus Question 3: How much is and how can we conceptualize one million? One billion?

Getting a handle on numbers as large as millions and billions is a challenge for humans, who measure lifespans in less than 100 years. Such large numbers are truly "far out" and require physical models and mathematical reasoning to visualize and conceptualize.

Metric Magnitudes, Nested Numbers, and Baffling Boxes: Display, "dissect," and discuss a sequence of metric boxes as a visual model for increasing powers of 10 from the small to the large:

- 1 crystal of table salt ~ $1mm^3$ (look at this with a hand lens or handheld microscope if time permits)
- $1,000 \ mm^3 = 1 \ cm^3$
- $1,000 \ cm^3 = 1 \ dm^3 = 1 \ L = 10^6 \ mm^3$
- $1,000 \ dm^3 = 1 \ m^3 = 10^6 \ cm^3 = 10^9 \ mm^3$

See *More Brain-Powered Science*, Activity #17 for details on assembling and using a cubic meter with dowel rods (or simply wedge 12 metersticks into 8 corner pieces of Styrofoam). Challenge learners to come up with their own representations for these large numbers using other concrete models with common school objects (e.g., standard-count boxes of paper clips or thumbtacks). For creative inspiration, consider projecting images from children's books such as

- *How Much Is a Million,* by D. M. Schwartz and S. Kellogg (1985). This book contains a number of fun, visual representations and analogies for large numbers (e.g., a fish bowl big enough to hold a million goldfish would be large enough to hold a 60-foot whale while a billion goldfish would need a bowl as big as a sports stadium).

- *Big Numbers and Pictures That Show Just How Big They Are,* by E. Packard and S. Murdocca (2000). *Note:* 3-D models that make use of volume to contain smaller objects enable one to display very large numbers in relatively small spaces; this is both an advantage and a disadvantage in terms of getting learners to visualize power-of-10 magnitudes. The Explore phase uses several linear models that complement this more compact 3-D one.

Focus Question 4: Has the Earth always contained a biosphere with living organisms? How has the diversity and specific mix of organisms varied over Earth's history? How did early life forms transform their environment and "set the stage" for subsequent life forms? How does a diversity of life forms make for more stable ecosystems? Were the major worldwide, multispecies extinctions of the past a "bad thing" from the perspective of humans? How should we respond to current human-induced species extinctions? What evidence and argument do scientists use to address questions such as these?

These broad, open-ended questions are designed to set the stage for the Explore phase and for life science classrooms, a subsequent 5E Teaching Cycle that would focus on the biological evidence for the evolution of life (e.g., anatomical, behavioral, biochemical, genetic, paleontological, and physiological patterns). Elicit an open-ended discussion and note the learners' perspectives without weighing in with the correct answers. Share the Murray Gell-Mann quote to emphasize that science itself (including our understanding of evolution) evolves.

2. *Explore:*

a. *Hands-On Exploration: Earth Events Evolve (Relative Geobiological Timeline):* Randomly distribute to each learner one (of the 34) 5 in. × 8 in. card with one Earth event per card (without the numeric data) from the abbreviated list provided in the Answers

to Questions in Procedures section. Ask learners to read their cards and mix with their peers to form a relative timeline (e.g., along the classroom's diagonal) that they think captures a probable sequence from the oldest to the most recent events in Earth's ongoing geobiological "story." Present this paleontological, historical storytelling task as analogous to organizing the chapters in a randomly arranged book or multiple books in a popular series such as *Harry Potter* (but be sure to note that the real fossil record is not randomly arranged). As a series of separate line segments develops, learners should interact with those on their left (designated for older events) and right (designated for more recent events) to discuss possible logical arguments as to why their sequence of three or more events "makes sense" (or not). The teacher can challenge triads or longer sequences of learners to reconsider their positions if desired, but do not be concerned with teaching or telling all the "right" answers at this time. Focus on challenging learners to begin to explore the logic of evolution (or descent with modification) in light of the interaction of geology and biology. Make note of misconceptions that are likely to arise in the learners' conversations. Poll the class to see how long they believe Earth contained only microbes. Briefly discuss the notion of decoupling Earth and human history by labeling 4.55 billion years ago as "time zero" for our planet, rather than the standard method of dating Earth's events in millions (or billions) of years ago (mya) from our current time.

After making a written record of the learner-generated timeline (or simply taping their cards to the wall), use actual dates for Earth's evolving natural history to rearrange their original sequence as necessary. Though this information is provided to learners as "received knowledge," it is important to emphasize again the concurrence of views arrived at through multiple scientific disciplines and to challenge learners to consider if the actual dated sequence of interrelated events concurs with or "makes more sense" than the sequence they developed as a class. *Note:* Studying the logic of the continuity and evolution of extinct and extant species of life (i.e., prokaryotes to eukaryotes to multicellular and macroscopic eukaroytes) should be a yearlong theme in life science courses.

b. *Hands-on Exploration:"Wiping Clean" Misconceptions About Evolutionary Time:*

Focus Question 1: *Counting Cards and Counting Time:* How does the linear arrangement of evenly spaced cards look like time-lines in history textbooks? How does the card sequence grossly misrepresent Earth's history?

Briefly discuss the idea of an absolute (versus relative) time and the unequal, nonlinear "distances" (i.e., number of years) between different pairs of cards. Geological time charts in books and posters typically represent this discontinuity with a broken or wavy timeline. With some flair and sense of fun, introduce the "Toilet Paper Timeline of Science" as an inexpensive, long roll of paper that can be used as a scale model for the "deep time" of Earth's history. The scale and calculations for particular events can be given to or derived by the learners depending on their experience and the amount of instructional time allotted. If learners are asked to do the mathematical calculations associated with finding the location of their original Earth event card on the toilet paper timeline, have them check their classmates' calculations prior to unrolling the toilet paper down a long hallway and positioning themselves (and their event cards) at the appropriate places alongside the toilet paper. Many common brands of toilet paper have 280 sheets per roll and will easily fit in typical school hallways. Some cheaper, single-ply brands of toilet paper come with 1,000 sheets.

SCALE: 280 sheets/roll × 11.1 cm/sheet = 3,108 cm = 31.08 m
4.55×10^9 yrs /280 sheets = $4,550 \times 10^6/280$ = 16.25×10^6 yrs/sheet
16.25×10^6 yrs/11.1 cm long sheet = 1.46 million yrs/1 cm or 1 million yrs/0.68 cm

Focus Question 2: *Toilet Paper Timeline:* How does the history of living organisms (bacteria to humans) compare in length (duration) to Earth's history? How many sheets of toilet paper would this be from Earth time zero? Where would other key significant Earth events occur on the Toilet Paper Timeline? After the event cards are positioned alongside the toilet paper, discuss their relative positions and have all the learners "take a hike" along the historical timeline to develop both a visual and bodily kinesthetic perspective on the scale of time involved. Depending on the

prior background of the learners, they can be asked to provide a guided tour commentary as the class stops at their particular event.

Focus Question 3: *Time Travel Analogy:* Assume a year of time is represented by one mile and that you could travel back in time in a conventional car traveling at the speed limit of 65 mph (or use 100 mph for easier calculations if desired). How long would your auto trip back to see the beginning of life on Earth take? How many typical human lifetimes is this? How many generations?

Focus Question 4: *Bacteria: Little in Size, But Big in Numbers and Old in Years (as Species):* Bacteria have always been and still are the most numerous life form on Earth (i.e., the number of bacteria in and on macroscopic organisms can exceed the number of eukaryotic cells the organisms contain). Assuming bacteria can reproduce every 20 minutes, what's the maximum number of generations of bacteria that could have theoretically existed over the past 3.8 billion years?

Depending on whether this 5E Teaching Cycle is used with Earth or life science classes, additional Explore-phase laboratory activities and simulations are available from McComas (1994), the National Academy of Sciences (1998, p. 9), Tarbuck (2003), and a number of sites listed in the Resource Books and Articles and Internet Connections for Teaching Evolution on pages 214–216 and Internet Connections listed below. A highly detailed, full-color Correlated History of Earth poster is available from Pan Terra, Inc. (WMPT001; $25 from the Worldwide Museum of Natural History; 605-574-4760 or *www.wmnh.com*).

3. *Explain:* Use select readings from textbooks and popular science magazines, computer and people-as-actor-type simulations, videos, virtual field trips to museums (see Internet Connections), and interactive, multimedia-enriched lectures and discussions spread over multiple days to develop an understanding of the chain of empirical evidence and logical arguments that support the idea of "deep time" as related to concepts from the following areas:

• *Earth Sciences*, such as sedimentation and relative dating via biostratigraphy, absolute dating via radiometric techniques (based on the constant half-lives of various naturally occurring radioisotopes), the improbability of fossil formation and "transitional gaps," and how paleontologists "read" the fossil record to "tell a story" that links key geological, chemical, and biological events in Earth's history (e.g., plate tectonics and the changing composition of gases in the atmosphere as related to biochemical metabolic processes).

- *Biological Sciences*, such as comparative anatomy, homologous structures, and adaptive radiation that supports the claim that current species evolved via descent with modification from ancestral forms, some of which are preserved in the fossil record over long expanses of time.

- *Nature of Science*, such as how skeptical review requires theories to have both explanatory power with respect to previously acquired evidence and predictive power with respect to guiding future research. Be sure to note that "empirical evidence" does not necessarily imply controlled experimental research. Research that "predicts" observational discoveries of "new" fossils, biochemical lineage markers, and other evidence of "nature's experiments" already carried out over Earth's immense history without human intervention counts as "real science."

The key to the Explain phase is to challenge learners to use minds-on skeptical reasoning to process previous hands-on activities and additional science concepts introduced by the teacher so that they understand (versus mindlessly "believe") the idea of deep time and how it is central to both geology and biology. A resource for high school classrooms that emphasizes the life sciences part of the story of evolution is the seven-part, eight-hour PBS TV series *Evolution: A Journey Into Where We're From and Where We're Going.* The DVDs, a teacher's guide, an extensive website, educational outreach initiatives, and a HarperCollins companion book are available from WGBH Boston Video (800-949-8670) for $19.95 per episode or $99.95 for the four-DVD set (WG35469; *www.pbs.org/wgbh/evolution/shop/index.html*). Show #2, "Great Transformations," and #3, "Extinctions" (57 minutes each), are especially useful for this 5E. A resource that is more appropriate for middle-level students are the *Bill Nye the Science Guy* episodes "Dinosaurs," "Evolution," and

"Fossils." These three titles are available for purchase from Disney Educational Productions ($29.99 for 26-min. DVD; 800-295-5010; *http://dep.disney.go.com*).

4. *Elaborate:* Formative assessment is ongoing throughout the previous activities. If the toilet paper (or equivalent-length roll of more durable paper) timeline is taped to the walls in the hall, learners can add relevant artwork and information above and below the timeline for a temporary display to catalyze out-of-class (even schoolwide) conversations about expanding the idea of "history" to include the huge expanses of time before the arrival of humans (much less the much shorter time frame of written human history). Depending on whether this 5E Teaching Cycle is being used in an Earth or life science classroom or as an inter- or cross-disciplinary unit, a variety of real-world, challenge-type questions or projects can be used.

   a. *Earth Sciences:* On August 3, 1960, geologists who were exploring a sandstone cliff in Spitsbergen found 13, 30 in. long footprints of *Iguanodon*. This large dinosaur species died out during the Cretaceous period some 100 million years ago. Spitsbergen lies well within the Arctic Circle. Why did this find pose a problem, and how was this seeming discrepancy explained? *Note:* Other examples of seemingly oddly located (if one views the Earth's continents as fixed in location over long periods of time) or matched fossils can be researched.

   b. *Biological Sciences:* Does evolution always require millions of years to occur? Can evolutionary changes ever be noticed in human time frames? The treatment of microbial diseases is an "evolutionary race" with an "enemy" who relies on time-accelerated rates of natural selection (i.e., relative to most macroscopic species). What factors account for the ability of germs to "adapt" to our best weapons (i.e., antibiotics as "wonder drugs")? A similar phenomenon is observed in the case of pesticide-resistant insects that destroy agricultural crops worldwide? What is or are the possible human responses to these two examples of rapid biological adaptations?

   c. *Mathematics of Scale Models:* What scale would science textbook publishers need to use to represent 4.55 billion years of Earth history if they placed a continuous timeline along the bottom of each of the textbook's 600, 20 cm long pages? Alternatively, have learners watch and check the mathematics in a video clip from the end of Episode 1 of the *Cosmos* series, where Carl Sagan presents the entire 15–20 billion year history (including Earth's 4.55 billion years) of the cosmos reduced to a 365-day Earth calendar year (see Internet Connections for video and data tables).

d. *Fossil Footprints: Reading Clues and Science Storytelling:* See Activity #5, Extension #1, and the Murray Gell-Mann Points to Ponder quote to highlight the tentative nature of science.

e. *Science-Technology-Society (STS) Issues:*

(1) *Creationism, Intelligent Design (ID), and the Nature of Science:* If curricular time, the maturity level of learners, and collaborations with English and/or social studies teachers allow, consider exploring the broader interdisciplinary issues connected with the misconstrued and ill-constructed "evolution (atheism) versus creationism" debate. It is important to note that although the theory, implications, and applications of evolution provide a lively area of ongoing research, this particular debate is neither a case of constructive competition for survival between two rival *scientific* theories that deserve equal time in a *science* classroom nor a case of science against a single monolithic, monotheistic religion. Instead, it is more akin to a political struggle where misinformation, distortion of facts, and the most extreme minority views from both sides tend to capture most of the media's attention (i.e., "young Earth" biblically bound creationists versus evangelical atheistic scientists who argue the theory does indeed make the idea of a god superfluous at best and pernicious at worst). Sample resources that are listed on pages 214–216 include articles (e.g., Moore 1998–1999; Rudolph and Stewart 1998), books (e.g., Miller 1998; National Academy of Sciences 1998, 1999, 2008; Skehan and Nelson 2000), and internet sites. See also the following sources:

- The previously mentioned PBS *Evolution* series: Show #1, ch. 8, "God"; 11:31 min.) and ch. 9, "A Scientist Discusses Religion" (4:53 min.); Show #7, "What About God" (57 min.)
- The *NOVA* episode "Judgment Day: Intelligent Design on Trial" and the associated website (below)
- *Star Trek: Voyager's* "Distant Origin" episode

The latter science fiction story creatively blends a Galileo-heresy-type inquisition hearing with a Darwinian-type context where an intelligent, scientifically advanced reptilian species is horrified to discover its evolutionary link to an ancient, distant Earth that later spawned an intelligent (but less so) mammalian species of humans with whom they now share many genetic markers. A wide variety of science fiction shows and movies (e.g., *Twilight Zone, Planet of the Apes, Legends,* etc.) have also featured evolutionary themes and can be analyzed (as cross-disciplinary projects) in terms of the validity of their representations of the nature of science and specific concepts related to evolution. *NSTA Report's* Blick on Flicks column (see Internet Connections) provides a nice model for this kind of analysis.

(2) *Evolution Matters to Everyone: Agriculture, Genetic Engineering, Medicine, and More*: Learners can research how insights from "deep" human history (i.e., our more than 100,000 years as *Homo sapiens*) offer insights into improvements in both preventative (e.g., healthy diet, physical exercise, social networks, vaccinations, and genetic screenings) and ameliorative medicine (e.g., antibiotics and gene therapies) and modern, more environmentally friendly agriculture. See UC Berkeley's Understanding Evolution (search: How does evolution impact my life?) website for information about these two applications and many others. The book *Why We Get Sick: The New Science of Darwinian Medicine,* by R. M. Nesse, MD, and G. C. Williams, PhD (1994) is a readable resource on evolution-informed medicine.

5. *Evaluate:* Learners can be challenged to develop and/or critique other visual analogical models such as 365 days per year, 30 days per month, or 24 hours per day equaling Earth's 4.55-billion-year timescale (see Resource Books and Articles: Packard 1994; Skehan and Nelson 2000; Internet Connections sites such as ASM, ASU, Cosmic Calendar, Deep Time, Evolution Timeline, and Kentucky Geological Survey). Other summative projects might include evolution-related advertising campaigns, bumper stickers, cartoons, computer simulations, debates, displays, fictional news releases, historical skits, and songs that confront misconceptions and promote dialogue and understanding. Summative evaluation for a larger unit on evolution should not be limited to only standard paper-and-pencil tests of "regurgitated facts," but instead should allow learners to demonstrate higher levels of understanding through a variety of types of assessment that include attention to both "how we know what we know" and real-world applications of evolutionary science. That said, well-designed paper-and-pencil tests do allow for a fairly rapid assessment of student understanding of a range of important ideas. Consider the following items (with bold-printed correct answers) as examples (also see Internet Connections: San Diego State University).

1. Extinctions are a rather rare phenomenon in the history of our planet and have largely been driven by humans.	T	**F**
2. Organisms exist today in essentially the same form in which they always have.	T	**F**
3. Evolutionary theory is supported by historical and ongoing investigations.	**T**	F
4. Most of Earth's organisms came into existence at about the same time.	T	**F**
5. The age of the Earth is about 4.6 billion years.	**T**	F
6. All living organisms ($10–100 \times 10^6$ species) are related to each other.	**T**	F
7. The theory of evolution is based on speculation and not valid scientific observation and testing.	T	**F**
8. The written record of human history is only a small fraction of *Homo sapien* history.	**T**	F

9. STATEMENT: Bacteria and insects can evolve more rapidly than mammals (e.g., humans) because ...
REASON: They have much higher reproduction rates and shorter generational times.

a. The statement is true, but the reason is false.	c. Both the statement and reason are false.
b. The statement is false, but the reason is true.	**d. Both the statement and reason are true.**

10. STATEMENT: The HIV virus that causes AIDS is capable of rapid evolution and as a result ...
CONCLUSION: Developing a vaccine that will have lasting effectiveness is difficult.

a. Statement is true, but the conclusion is false.	c. Both statement and conclusion are false.
b. Statement is false, but the conclusion is true.	**d. Both statement and conclusion are true.**

11. Penicillin, the first "wonder drug" antibiotic, is proving to be increasingly ineffective in treating many types of bacterial infections *because*

   a. penicillin produced changes in individual bacteria that made them increasingly more tolerant of the drug over time with repeated exposure.
   b. bacteria are capable of using part of the penicillin molecule as a food source.
   c. **pre-existing resistant strains of bacteria were selected by misuse and overuse of the drug and therefore have become more common.**
   d. cheaper mass production of generic penicillin has resulted in a lower-quality product.
   e. bacteria have mutated because of increased exposure to high-energy ultraviolet radiation due to the hole in our atmosphere's ozone layer.

Debriefing

When Working With Teachers

As time permits, discuss the analogy drawn between learning as conceptual change and evolution as described in the previous section on Science Education Concepts. The sequence of phases in a 5E Teaching Cycle (i.e., Engage, Explore, Explain, Elaborate, Evaluate) is designed to provide scaffolded experiential and conceptual support for learners' understanding of complex ideas to "evolve" across a given instructional unit and year. Multi–grade level *learning progressions* need to be developed, tested, and refined by teachers, researchers, and textbook authors to more efficiently support students' conceptual growth toward deeper understanding of the "big ideas" of science (NRC 2010a). The importance of explicitly teaching big ideas such as *constancy and change, scale, models,* and *systems* as K–12 themes is emphasized in multiple science reform documents (see Appendix C). These big ideas (and related core scientific theories) provide a broader conceptual framework (or "forest") that helps students remember, relate, and refine their understanding of the many individual concepts (or "trees") of science.

When Working with Students

Though teachers cannot be expected to single-handedly resolve the religious and political issues surrounding the misframed "evolution equals atheism versus creationism" debate, they can and should confront learners' motivational barriers and misconceptions about evolution and the nature of scientific inquiry. Core scientific concepts and theories such as evolution need to be taught (versus presented) with adequate attention to how

- analogies, models, and visual representations make otherwise abstract, counterintuitive, and difficult to perceive and conceive ideas more "sensible" (i.e., experientially accessible and logically defensible). See, for example, Gilbert and Watt Ireton (2003); Hackney and Wandersee (2002); Harrison and Coll (2008); and Hoagland and Dodson (1998).
- developing student' sense of "how we know what we know," including especially the interconnections with other scientific concepts and theories, helps them construct a tightly interwoven tapestry of meaning. Epistemology is a critical component of the nature of science and scientific literacy.
- featuring the personal applications and social relevance of scientific theories helps students see how they are both parsimonious and productive (i.e., have both broad explanatory and predictive power) in the "real world."
- catalyzing conceptual change and meaningful, lasting learning takes time. Engaging learners in a dialogue of discovery that invites them to question the answers takes more time than simply providing them authoritative, "just the facts" answers via a monologue of conclusions approach. Individual conceptual change is an evolving, dialectical process, as is the ongoing evolutionary development of scientific theories.

Without these components, science instruction can become a form of indoctrination where learners are expected to quickly and mindlessly accept (and subsequently regurgitate) the "answers" for tests based on the unquestioned authority of the textbook and teacher. This kind of "belief" in science is diametrically opposed to the nature of science, and as such is poor preparation for both a scientifically literate citizenry and future scientists. There are two

additional factors to consider that are specific to this particular 5E Teaching Cycle:

- Learners need multiple rationally sequenced experiences that draw on a full array of intelligences (Armstrong 2000; Gardner 1999) to counter their everyday conceptions of "history" and its unduly time-myopic association with humans and human life spans.
- Some learners will find it difficult to accept (avoid the term *believe*) the scientific theory of evolution, even if they understand it conceptually, due to deeply held, conflicting personal religious beliefs. Though religious beliefs should be respected, it is important that all students understand that science in general, and evolution theory in particular, does not require one to be an atheist. Rather, science is intentionally agnostic (a term coined by Thomas Henry Huxley) in its methodological approaches to develop provisionally accepted, proximate-cause-type, naturalistic explanations of natural phenomena (i.e., answers to "what and how" questions in terms of quantifiable amounts of matter and energy). Science does not and cannot directly address the domain of potential supernatural causes that are the primary focus of religion ("ultimate whys") except in the context of debunking pseudoscientific claims (see *More Brain-Powered Science,* Activity #5, and the previously mentioned "creation science").

Evolution, properly addressed as a cross-disciplinary, unifying theme that pervades an entire course of study, provides a perfect opportunity to engage learners in critically analyzing the nature of scientific inquiry and confronting commonly held myths about science, such as the meaning of the term *theory* (McComas 1996; Rudolph and Stewart 1998; see Resource Books). The mistaken separateness of the numerous topics and chapters within textbooks belies the underlying unity that powerful, interlinked theories provide to science and science education (Duschl 1990; see Resource Books). Science as an evolving product and process is inherently constructivist and parsimonious in its efforts to explain the broadest range of phenomena in terms of the fewest number of basic assumptions, unifying concepts and principles, and empirically supported laws and theories. The ever-evolving theory of evolution makes a perfect case study for this point.

Resource Books and Articles for Teaching Evolution

- American Association for the Advancement of Science (AAAS). 2006. *Evolution on the front line: An abbreviated guide for teaching evolution* (25-page booklet). *www.project2061.org/publications/guides/evolution.pdf*

- Biological Sciences Curriculum Study (BSCS). 2005. *The nature of science and the study of biological evolution.* Colorado Springs, CO: BSCS and Arlington, VA: NSTA Press. (book and CD-ROM Teacher Guide)

- Duschl, R. A. 1990. *Restructuring science education: The importance of theories and their development.* New York: Teachers College Press.

- Hoagland, M., and B. Dodson. 1995. *The way life works: Everything you need to know about the way all life grows, develops, reproduces, and gets along.* New York: Times Book/Random House. See especially Chapter 7, "Evolution."

- McComas, W. F. 1994. How long is a long time: Constructing a scale model of the development of life on Earth. In *Investigating evolutionary biology in the laboratory*, ed. W. F. McComas, pp. 31–39. Reston, VA: National Association of Biology Teachers.

- McComas, W. F. 1996. Ten myths of science: Reexamining what we think we know about the nature of science. *School Science and Mathematics* 96 (1): 10–16.

- Miller, J. B. 1998. *An evolving dialogue: Scientific, historical, philosophical, and theological perspectives on evolution.* Washington, DC: American Association for the Advancement of Science.

- Moore, R., ed. Eight-part series on "Creationism in the United States." *The American Biology Teacher* 60 (7)–61 (5).

- National Academy of Sciences (NAS). 1998. *Teaching about evolution and the nature of science.* Washington, DC: National Academies Press. *www.nap.edu/openbook.php?record_id=5787*.

- National Academy of Sciences (NAS). 1999. *Science and creationism: A view from the National Academy of Sciences.* 2nd ed. Washington, DC: National Academies Press. *www.nap.edu/catalog.php?record_id=6024*.

- National Academy of Sciences (NAS). 2008. *Science, evolution, and creationism*. Washington, DC: National Academies Press. *www.nap.edu/catalog.php?record_id=11876*.

- National Research Council (NRC). 1996. *National science education standards*. Washington, DC: National Academies Press. *www.nap.edu/catalog.php?record_id=4962*.

- Packard, E. 1994. *Imagining the universe: A visual journey*. New York: Perigee Books.

- Rudolph, J. L., and J. Stewart. 1998. Evolution and the nature of science: On the historical discord and its implication for education. *Journal of Research in Science Teaching* 35 (10): 1069–1089.

- Skehan, J. W., and C. E. Nelson. 2000. *The creation controversy and the science classroom*. Arlington, VA: NSTA Press.

- Tarbuck, E. J. 2003. *The theory of plate tectonics*. Albuquerque, NM: TASA Graphics Arts. *www.tasagraphicarts.com* ($59 for CD-ROM; 2 levels: grade 7–college).

- Whitfield, P. 1993. *From so simple a beginning: The book of evolution*. New York: Macmillan.

- Wilson, D. S. 2007. *Evolution for everyone: How Darwin's theory can change the way we think about our lives*. New York: Bantam Dell/Random House.

- Woodrow Wilson National Fellowship Foundation. 1995. *Evolution: A context for biology. www.woodrow.org/teachers/bi/1995/index.html*.

- Zimmer, C. 2001. *Evolution: The triumph of an idea*. New York: HarperCollins Publishers. Companion to the PBS series.

Internet Connections

Note: Given that evolution is a critical, yearlong theme ("big idea") in both the biological and Earth sciences, the following list is much longer than typical for this book. Several sites on creation science and intelligent design are included for teachers who wish to explore the "evolution versus religion debate" as an STS/interdisciplinary issue. These sites are identified with an asterisk (*).

- AAAS Dialogue on Science, Ethics and Religion: *www.aaas.org/ spp/dser*

- AAAS Project 2061: Evolution on the Front Line Resources (variety of links, news stories, position statements, and pamphlets): *www.aaas.org/news/press_room/evolution*

- Access Excellence (source for all biology topics): *www. accessexcellence.org*

- ActionBioscience: Issues in Evolution (several articles, including "Once in a Million Years: Teaching Geological Time"): *www. actionbioscience.org/evolution/index.html*

- American Geological Institute: *Evolution and the Fossil Record* (with Geological Timescale): *www.agiweb.org/news/evolution*

- American Museum of Natural History (New York): *www.amnh.org*

- American Society for Microbiology (ASM): Classroom Activities: Earth History: Time Flies No Matter What the Scale (cartoons timeline and geologic timescale mapped to a yearly calendar): *www.asm.org/index.php/education/classroom-activities.html*

- Annenberg Foundation's Teachers' Professional Development: *Rediscovering Biology* (animations, images, and QuickTime movies on evolution): *www.learner.org/channel/courses/biology/units/ humev/images.html*

- Arizona State University's (ASU) Institute of Human Origins: *http://iho.asu.edu*

    Becoming Human (interactive online timeline): *www. becominghuman.org*

- Awesome Science Teacher Resources by Nancy Clark: Evolution Activities, Labs and Links: *www.nclark.net/Evolution*

- BBC Prehistoric Life (games, simulations, and more related to evolution): *www.bbc.co.uk/sn/prehistoric_life/redesign.shtml*

- Biology Corner: Evolution and Taxonomy (labs, simulations, concept maps, analogies, etc.): *www.biologycorner.com/lesson-plans/ evolution-taxonomy*

- Biology in Motion: Evolution Lab (Simulation): *www. biologyinmotion.com*

- Breve (free, open-source software package for building 3-D simulations of multi-agent systems and artificial life): *www.spiderland.org*

- Concord Consortium: Biology Simulations: *www.concord.org/activities/subject/biology*

    Free download: Competition (MS), Conflicting Selection Pressures (HS), Modern Genetics (HS), Natural Selection (ES), Population Explosion (HS), Predators and Prey (ES), and Virtual Greenhouse (ES). See also the article "Teaching Evolution With Models": *www.concord.org/publications/newsletter/2009-spring/evolution.html*

- Contextual Science Teaching: *http://sci-ed.org.*

    See Centre for Science Stories:

    Biological Sciences (Alfred Russel Wallace and Charles Darwin): *http://science-stories.org/category/subject/biological-sciences*

    Earth Sciences (Age of the Earth: A Very Deep Question and Alfred Wegener/Continental Drift): *http://science-stories.org/category/subject/earth-sciences*

- Cosmic Calendar (from Carl Sagan's 1977 book *The Dragons of Eden* and featured in the *Cosmos* PBS series, Episode 1; cosmic chronology compressed to a single Earth year):

    http://visav.phys.uvic.ca/~babul/AstroCourses/P303/BB-slide.htm (data tables/charts)

    www.youtube.com/watch?v=g2qezQzfgIY (4:52 video from original *Cosmos* series)

    www.youtube.com/watch?v=igPPh8_bXWw (longer version; features evolution)

- Darwin Pond (Artificial Life Simulation): *www.ventrella.com/Darwin/darwin.html*

- DarylScience: Biology Demos: *www.darylscience.com/DemoBio.html*

    See demonstrations #1, "Opposable Thumbs," and #2, "Darwin's Finch Beaks

- Deep Time—the Geological Time Scale" (with a number of visual analogies, including a roll of toilet paper): *http://serc. carleton.edu/quantskills/methods/quantlit/DeepTime.html*

- Dinobase at the University of Bristol (pictures, data+): *http:// dinobase.gly.bris.ac.uk/index.htm*

- *Discovery Institute (conservative think tank: science and culture and intelligent design): *www.discovery.org*.

    See critique of supposed "flaws" in PBS *Evolution* series: *www. reviewevolution.com/index.php*

- Disney Educational Productions: *Bill Nye the Science Guy*: *Dinosaurs, Evolution, and Fossils* ($29.99/26 min. DVD): *http://dep. disney.go.com*

- Evolution and the Nature of Science Institute: *www.indiana. edu/~ensiweb/home.html*

- Evolution Timeline (Big Bang through the present animation): *www.johnkyrk.com/evolution.swf*

- EvoTutor Interactive Simulations: *www.evotutor.org*

- Geological Society of America: Geological Time Scale: *www. geosociety.org/science/timescale*

- Howard Hughes Medical Institute, *Biointeractive* (free DVDs of virtual labs, lectures, etc.): *www.hhmi.org/biointeractive*.

    See for example *Evolution: Constant Change and Common Threads* and *Evolution: Fossils, Genes and Mousetraps* DVDs.

- Human Origins and Evolution in Africa: *www.indiana. edu/~origins*

- *Institute for Creation Science (biblically based, anti-evolution): *www.icr.org*

- *Intelligent Design Resources: *www.intelligentdesign.org/resources.php*

- International Darwin Day Foundation: *www.darwinday.org*

- Kentucky Geological Survey: *It's About Time* (variety of visual analogies and resources): *www.uky.edu/KGS/education/about_ time.htm*

- Millerandlevine.com (resources on evolution and teaching controversy by textbook authors): *http://millerandlevine.com*
- Museum of the Earth and the Paleontological Research Institution (PRI): *www.museumoftheEarth.org*
- National Academy of Sciences and National Academies Press:

  Evolution Resources: *www.nationalacademies.org/evolution*

  Science and Creationism (booklet): *www.nap.edu/openbook. php?record_id=6024*

  Teaching About Evolution and the Nature of Science (1998 book that includes classroom activities such as Footprint Puzzle/ Fossil Tracks and Mystery Cube set in 5E Teaching Cycle): *http://books.nap.edu/openbook.php?isbn=0309063647*

  Science, Evolution, and Creationism (2008 book): *www.nap.edu/ catalog.php?record_id=11876*

- National Association of Biology Teachers: Position Statement on Teaching Evolution: *www.nabt.org/websites/institution/index.php?p=35*
- National Center for Science Education: Defending the Teaching of Evolution in Public Schools: *http://ncse.com*
- National Science Teachers Association: *NSTA Reports* Blick on Flicks column: *www.nsta.org/publications/blickonflicks.aspx*
- National Science Teachers Association: Evolution Resources (e.g., Position Statement, NSTA books, external links, etc.): *www. nsta.org/publications/evolution.aspx*

  NSTA Tool Kit for Teaching Evolution: SciLinks at *www.nsta.org/ publications/press/extras/evolutiontoolkit.aspx*

- NetLogo (free cross-platform, multi-agent programmable modeling and models library [evolution]): *http://ccl.northwestern.edu/netlogo*
- *Northwest Creation Network: "Creation Science"/Anti-Evolution Cartoons and Humor ("good" source of science and religion misconceptions): *http://nwcreation.net/humor.html*
- *NOVA: Intelligent Design on Trial* (DVD): *www.pbs.org/wgbh/nova/ evolution/intelligent-design-trial.html*
- Online Literature Library (*The Voyage of Beagle, Origin of Species,* and *Descent of Man*): *www.literature.org/authors/darwin-charles*

- Pangea Animations:

    http://en.wikipedia.org/wiki/Pangaea

    www.classzone.com/books/earth_science/terc/content/visualizations/es0806/es0806page01.cfm?chapter_no=visualization (with dates for the past 150 million years)

    www.ucmp.berkeley.edu/geology/tectonics.html

    http://education.sdsc.edu/optiputer/teachers/overview.html (no dates)

    http://apod.nasa.gov/apod/ap001002.html (Earth, 250 million years from now)

- PBS *Evolution* series and related resources: *www.pbs.org/wgbh/evolution*

    Deep Time/History of Life: *www.pbs.org/wgbh/evolution/library/03/index.html*

- PhET Interactive Simulations:

    Natural Selection: *http://phet.colorado.edu/en/simulation/natural-selection*

    Radioactive Dating Game: *http://phet.colorado.edu/en/simulation/radioactive-dating-game*

- Rev. Ken Collins website: Evolution Versus Creationism: A Stupid and Arrogant Fight: *www.kencollins.com/bible-i4.htm* (pro-science/pro-religion site with analogy)

- Richard Dawkins Foundation for Reason and Science (evolutionary biologist and atheist): *http://richarddawkins.net*

- Richard Milner's Darwin's Universe (scholarship, music, art, and entertainment): *http://darwinlive.com*

- Royal Tyrrell Museum ("Canada's authority of palaeontology"): *www.tyrrellmuseum.com*

- *Scientific American Digital Special Editions*: Our Ever-Changing Earth, Becoming Human, Dinosaurs, and a New Look at Human Evolution: *www.scientificamerican.com/special/toc.cfm?issueid=34&sc=singletopic*

- Serendip: Hands-on Activities for Teaching Biology to High School or Middle School Learners:

See Evolution by Natural Selection simulation and other resources: *http://serendip.brynmawr.edu/sci_edu/waldron*

- San Diego Museum of Man: *Footsteps Through Time: Four Million Years of Human Evolution*: *http://abouthumanevolution.net*

- San Diego State University: Biology Lessons for Prospective and Practicing Teachers: *www.biologylessons.sdsu.edu*

  Concept Cartoons for Evolution: *www.biologylessons.sdsu.edu/cartoons/concepts.html*

  Conceptual Inventory of Natural Selection (20-item misconceptions-based test): *www.sci.sdsu.edu/CRMSE/old_site/kfisher_conceptual.htm*

- Smithsonian Institution National Museum of Natural History: *www.mnh.si.edu*

  Department of Paleobiology: *http://paleobiology.si.edu*

- Sources of Science Cartoons (check for copyright and permission to use):

  Mark Anderson: *www.andertoons.com/cartoons/science*

  Ashleigh Brilliant (Brilliant Thoughts, Pot-Shots and more): *www.ashleighbrilliant.com*

  Nick Downes: *http://nickdownes.com* (*Big Science* and *Whatever Happened To "Eureka"?*)

  Benita Epstein: *www.benitaepstein.com*

  CartoonStock/science: *www.cartoonstock.com/directory/s/science.asp*

  Cartoonist Group: Science: *www.cartoonistgroup.com/bysubject*

  John Chase: Chasetoons Science Humor: *www.chasetoons.com/schum.html*

  Randy Glasbergen, *The Better Half* (and more): *www.glasbergen.com*

  Sidney Harris at Science Cartoons Plus: *www.sciencecartoonsplus.com* (number of cartoon books on science, education, medicine, psychology, etc.)

  Nick D. Kim, *Nearing Zero*: *www.lab-initio.com* and *www.nearing-zero.net/index.html*

Gary Larson, *The Far Side* (book series): *www.thefarside.com*

John McPherson, Close to Home: *www.physlink.com/Fun/McPherson.cfm* and *www.gocomics.com/closetohome*

Mark Parisi, *Off the Mark*: *www.offthemark.com/science/science.htm*

Science Humor—It's Alive (cartoons, video clips, and more): *www.sciencehumor.org*

Tom Swanson: *http://home.netcom.com/~swansont/index.html*

Bob Thaves, *Frank and Earnest*: *www.frankandernest.com*

- Talk Origins Archive: Exploring the Creation/Evolution Controversy: *www.talkorigins.org*

- Teachers' Domain: Digital Media for the Classroom and Professional Development:

  Search topic: *Evolution* yields more than 200 lessons with linked video clips; for example, see Fossil Evidence for Evolution (2–3 days, grades 9–12 lessons with video clips): *www.teachersdomain.org/resource/tdc02.sci.life.evo.lp_fossilevid*

- The Paleontological Society: Position Statement: Evolution: *www.paleosoc.org/evolutioncomplete.htm*

- University of California–Berkeley, Museum of Paleontology: *http://ucmp.berkeley.edu/index.html*

  Understanding Evolution (extensive teacher resources): *http://evolution.berkeley.edu*

  Geological Time (see Plate Tectonics and Web Geological Time Machine: *www.ucmp.berkeley.edu/exhibit/geology.html*

- U.S. Geological Survey:

  Dinosaurs: Fact and Fiction: *http://pubs.usgs.gov/gip/dinosaurs*

  Fossils, Rocks and Time: *http://pubs.usgs.gov/gip/fossils*

  Geological Time (online edition): *http://pubs.usgs.gov/gip/geotime*

- Virtual Fossil Museum (geological history, paleobiology, phylogenetics, and evolutionary biology information): *www.fossilmuseum.net*

- Wikipedia: *http://en.wikipedia.org/wiki*

Search for films: *Flock of Dodos: The Evolution-Intelligent Design Circus* and *Expelled: No Intelligence Allowed* for contrasting perspectives.

- Woodrow Wilson National Fellowship Foundation. *Evolution: A Context for Biology*: *www.woodrow.org/teachers/bi/1995.* Also available at *www.accessexcellence.org/AE/AEPC/WWC/1995*

- Worldwide Museum of Natural History (photos and educational products): *www.wmnh.com*

- YouTube: Video clips of widely varying scientific quality. See, for example,

    Evolution Is a Blind Watchmaker (~10 min. anti-intelligent design simulation): *www.youtube.com/watch?v=mcAq9bmCeR0*

    In the Year 2525/Exordium and Terminus (song and slide shows): *www.youtube.com/watch?v=TJU_UutAME4&feature=related*

    www.youtube.com/watch?v=5tLTb4P1HD8&feature=related

    www.youtube.com/watch?v=1FgSmdfRUus (with lyrics)

Answers to Questions in Procedure, steps #1, #2, and #4

1. *Engage:*

Focus Questions 1 and 2: Individual human life (10^2), written records of human history (10^3–10^4), *Homo sapien* history (10^5), history of life on Earth (10^6) → the age of the Earth (10^9). Paleontologists "read" the fossil record of life that dates back to microbes some 3.8 billion years ago. Geologists use radiometric dating techniques to date Earth back to 4.5–4.6 billion years ago.

2. *Explore:*

   a. *Note:* Given that radiometric dating techniques for calculating distant events in Earth history do not have anything close to the precision of human birthdays (i.e., year, month, day, hour, minute), the precision of the calculations based on these numbers is significantly less than the number of digits generated by the spreadsheet for the second and third columns on the following table. Measuring and placing specific events within the 11.1 cm length of a segment of toilet paper (TP) becomes especially challenging (and less precise) for the last few events on the following table. Emphasize the big picture presented by the visual analogy of deep time and what it says about the relative lengths of Earth, life, and human history.

Event	Age (millions of years ago)	Percentage of time	Number of sheets of toilet paper
Earth formed	4550	0	0
Oldest rocks	3800	16.48351648	46.15384615
First life (stromatolites/prokaryotes)	3500	23.07692308	64.61538462
Atmospheric O_2 from cyanobacteria	2000	56.04395604	156.9230769
First organized cells (eukaryotes)	1000	78.02197802	218.4615385
First multicelled animals	680	85.05494505	238.1538462
Cambrian "explosion" of life	570	87.47252747	244.9230769
First vertebrate animals	450	90.10989011	252.3076923
First land plants	430	90.54945055	253.5384615
First fish	400	91.20879121	255.3846154
First amphibians	365	91.97802198	257.5384615
First insects	350	92.30769231	258.4615385
First reptiles	320	92.96703297	260.3076923
First conifer trees	300	93.40659341	261.5384615
Extinction of trilobites	285	93.73626374	262.4615385
First mammals (small)	200	95.6043956	267.6923077
First dinosaurs	200	95.6043956	267.6923077
Opening of the proto-Atlantic	200	95.6043956	267.6923077
First birds	160	96.48351648	270.1538462
Breakup of Earth's supercontinent	150	96.7032967	270.7692308
Opening of North Atlantic	120	97.36263736	272.6153846
Opening of South Atlantic	92	97.97802198	274.3384615
First primates	80	98.24175824	275.0769231
Extinction of the dinosaurs	65	98.57142857	276
Collision of India with Asia	65	98.57142857	276
Antarctica splits from Australia	53	98.83516484	276.7384615

Event	Age (millions of years ago)	Percentage of time	Number of sheets of toilet paper
First horses	26	99.42857143	278.4
First apes	25	99.45054945	278.4615385
Homo habilis9p2.45erectus out of Africa	2	99.95604396	279.8769231
Homo sapiens neanderthalis	0.2	99.9956044	279.9876923
Homo sapiens sapiens	0.1	99.9978022	279.9938462
Historical written records (Sumeria)	0.005	99.99989011	279.9996923
Establishment of the United States	0.00022	99.99999516	279.9999865
Number of years of schooling (K–PhD)	0.00002	99.99999956	279.9999988

b. **Focus Question 2** *(Toilet Paper Timeline):* Microbial life (prokaryotes) began approximately 3.8×10^9 years ago or 0.75×10^9 years from Earth "time zero": 0.75×10^9 years$/4.55 \times 10^9$ yrs $= 0.1648$ (i.e., during the first 16% of Earth's history, there was no life present, or conversely, life has been present for 84% of Earth's history). For the first 0.1648×280 sheets $= 46.15$ sheets from Earth "time zero," there does not appear to have been any life. Until recently, it was believed that prokaryotes were the only forms of life for roughly the next two billion years. Research now seems to indicate that eukaryotes may have evolved as early as 2.7 billion years ago. A few representative calculations (based on a 280-sheet roll of toilet paper) include the following:

Cambrian "Explosion": 570 million years ago = 0.57×10^9 years: segmented worms, coral anemones, jellyfish, sponges, lamp shells, clams, snail, squid, trilobites, crabs, etc.

4.55×10^9 years – 0.57×10^9 years = 3.98×10^9 years from Earth "time zero"

$3.98/4.55 = 0.8747$ (87% of Earth history occurred *before* invertebrates)

0.8747 (280 sheets) = 245 sheets from Earth "time zero"

First "Homo": 2 million years ago: *Homo habilis – Homo erectus*

$4,550 \times 10^6$ yrs – 2×10^6 years = $4,548 \times 10^6$ years from Earth "time zero"

$4548/4550 = 0.99956$

0.99956 (280 sheets) = 279.88 sheets from Earth "time zero"

0.88 sheets (11.1 cm/sheet) = 9.8 cm from start of last sheet (#280) on roll

Homo sapiens neanderthalis: 200,000 years ago = 0.2×10^6 years
4550×10^6 years – 0.2×10^6 years = 4549.8×10^6 years from Earth "time zero"
$4549.8/4550 = 0.999960$
0.999960 (280 sheets) = 279.9877 sheets from Earth "time zero"
0.987 sheets (11.1 cm/sheet) = 10.96 cm from start of last sheet (#280) on roll

Homo sapiens sapiens, the sole "survivors" of the Homo line, date to approximately 100,000 years ago, overlapping in time and location (and perhaps interbreeding) with the Neanderthals, whom they subsequently replaced and displaced or partially assimilated. Modern humans show up in the last 0.07 cm (or 0.7 mm) of the toilet paper timeline, and all of written history is approximately one tenth of that final segment. Take some time to emphasize this startling perspective and the implications of our roles as intelligent and compassionate stewards for Spaceship Earth.

Focus Question 3 *(Time Travel Analogy):*

3.8×10^9 miles × 1 hour/65 miles × 1 year/8,760 hours = 6,674 years
Note: Earliest human cities were built approximately 6,000 years ago.

6,674 years/80-year lifespan = more than 83 lifetimes of auto-speed travel time
6,674 years/20-year reproduction cycle = approximately 334 generations

Focus Question 4 *(Bacteria: Little in Size, But Big in Number):*

3.8×10^9 years × 8760 hours/1 year × 60 minutes/1 hour × 1 generation/20 minutes = 10^{13} generations

4. *Elaborate:*

   a. *Earth Science:* With few exceptions, all dinosaur species were cold-blooded reptiles that lived in tropical or subtropical regions. Large, herbivorous dinosaurs required much larger amounts of vegetation than could be found in an Arctic wasteland. The "obvious" answer is that Spitsbergen could not have had a frigid climate at the time the *Iguanodon* herds

roamed. In fact, geological evidence for continental drift and plate tectonics suggests that most of today's continents were part of the supercontinent of Pangaea and later drifted to their present locations. See the Pangea animations in the Internet Connections.

b. *Biological Sciences:* The highly variable rates of evolutionary change in different species are related to the time between generations and reproduction rates (i.e., number of offspring over time). Selective advantages and fortuitous mutations in bacteria are "accelerated" because under optimal conditions, each single-celled bacteria can asexually reproduce and divide (into two fully functioning cells) in as little as 20 minutes, which can result in the exponential growth of subsets of the original population. Misused (e.g., prescriptions for viral infections) and overused antibiotics (e.g., excessive amounts in animal feed) "select for" the rise of antibiotic-resistant strains. Proper, more limited, and targeted use of existing antibiotics and development of new ones are critical to sustainable public health. Similarly, the fast reproductive cycle of some insects (e.g., often measured in days or weeks) and the large number of offspring generated by their sexual reproduction allow for a similar selection for pesticide-resistant insects. Use of integrated pest-management systems that creatively combine targeted use of chemical pesticides, natural predators (including microbes), and altered agricultural planting cycles can limit the rise of pesticide-resistant insects. In both cases, humans need to understand the idea of relative time when it comes to reproduction and our response to natural cycles and evolution of particular variants within species. Genetic engineering also opens up the possibility of intentionally accelerated human-designed evolution of "desirable" bacteria and viruses and genetically improved plants and animals (as compared to the slower, 10,000-year-old process of conventional, agricultural selective breeding).

c. *Mathematics of Scale Models*: 600, 20 cm wide pages = 12,000 cm = 120 m. This is nearly four times the length of the 280-sheet roll of toilet paper (4.55×10^9 yrs/12,000 cm = 3.79×10^5 yrs/cm). If textbooks did this, it would both make a very graphic visualization of the Earth's age and the history of life and a statement about the importance of evolution as a yearlong theme that ties together all the various topics, rather than merely as an isolated unit.

Activity 13

5 E(Z) Steps to Earth-Moon Scaling
Measurements and Magnitudes Matter

Expected Outcome*

Learners are surprised to learn that most textbook illustrations incorrectly represent the relative sizes and/or distance between Earth and its single moon. Although these and other visual representations of our solar system "lie" or grossly misrepresent how truly far out planetary objects are from one another, scale models can help modify textbook-induced misconceptions.

* This activity has been modified from the following source:
O'Brien, T., and D. Seager. 2000. 5 E(z) steps to teaching Earth-Moon scaling: An interdisciplinary mathematics/science/technology mini-unit. *School Science and Mathematics* 100 (7): 390–395.

Science Concepts

The AAAS *Benchmarks for Science Literacy* (1993) common themes of systems, models, and scale (see Appendix C) are all featured in this 5E mini unit that focuses on the relative diameters of and mean distance between the two bodies in our Earth-Moon system. Related mathematical concepts and skills include the mathematical design, construction, and use of scale models, critical thinking, estimation, measurement (linear and cubic), prediction, proportionality, and ratios.

Science Education Concepts

The concept of scale with respect to the relative sizes of and distances between Earth, its Moon, and other bodies within our solar system poses a unique curriculum-instruction-assessment (CIA) challenge for the following reasons:

- The relevant linear measurements fall outside students' normal "comfort zone" of everyday familiarity (i.e., 10^{-3} to 10^3 m or 1 mm through 1 km).
- The limitations-imposed human perceptual biases, Earth-bound observations, and visual misrepresentations in textbooks, posters, and models generate tenacious misconceptions that don't allow "mental space" for new scientifically valid ideas to take hold.
- Textbooks typically cover these concepts too quickly to uncover students' misconceptions and help them discover the underlying interrelated mathematical and scientific concepts and principles.
- Some teachers assume that the key ideas of astronomy can *only* be taught at night and require expensive telescopes.
- Today's students are decades removed from the days when it was either a science fiction fantasy or a national goal for humans to leave the gravitational limits of Earth and venture into outer space, and as a result, extraordinary solar system statistics now seem mundane (to the under- or miseducated).

Exploring students' alternative conceptions and catalyzing curiosity about the Earth and Moon system provides an opportunity to

- emphasize the history and nature of science and its reliance on empirical criteria, logical argument, and skeptical review (i.e., the use of visual-spatial, physical models as exploratory and explanatory tools);
- integrate science with mathematics (as well as history if desired); and
- help learners begin to appreciate the awe-inspiring scale of our solar system and the broader, truly far-out universe that makes all astronomical bodies analogous to cosmic islands separated by vast oceans of empty space.

As a *visual participatory analogy for science teacher education*, this 5E mini unit shows how an "intelligently designed" sequence of fun CIA activities (e.g., sports balls as scale models of planetary objects) provides cognitive "scaffolding" for the "hard construction work" of conceptual change. The CIA-assisted learning of science can be viewed as "intelligently designed and directed selection forces" that help students' mental models evolve from somewhat flawed, limited applicability, "primitive" ideas (i.e., misconceptions) to more up-to-date, scientifically "fit" ones. Both the practice and learning of science are FUNdaMENTAL processes that depend on the creative integration of hands-on and minds-on activity and the interaction between learners and a specific learning environment that "selects for" certain scientific habits of mind, skills, and concepts. Minds-on science teaching (MOST) catalyzes the MOST learning by relying on instructional processes that reflect the nature of science.

Materials

- Video clips of songs and movies cited in the Internet Connections
- An assortment of regulation-size sports balls (see 10 varieties listed in Table 13.1 on p. 242) in a large black garbage bag
- Metric measuring rulers or tape
- String or yarn
- Science posters and/or models of the Earth and Moon (see Internet Connections for sources)

Points to Ponder

When I heard the learn'd astronomer; When the proofs, the figures, were ranged in columns before me; When I was shown the charts and the diagrams, to add, divide, and measure them; When I, sitting, heard the astronomer, where he lectured with much applause in the lecture-room, How soon, unaccountable, I became tired and sick; Till rising and gliding out, I wander'd off by myself, In the mystical moist night-air, and from time to time, Look'd up in perfect silence at the stars.

—Walt Whitman (1819–1892), *Leaves of Grass* poetry collection

We came all this way to explore the moon, and the most important thing is that we discovered the Earth.

—William Anders (1933–), Apollo 8 astronaut who took the famous December 24, 1968, photo of "Earthrise" over the lunar horizon

It suddenly struck me that that tiny pea, pretty and blue, was the Earth. I put up my thumb and shut one eye, and my thumb blotted out the planet Earth. I didn't feel like a giant. I felt very, very small … That's one small step for [a] man; one giant leap for mankind … Here men from the planet Earth first set foot upon the Moon. July 1969 AD. We came in peace for all mankind.

—Neil Armstrong (1930–), Apollo 11 astronaut, first man to stand on the Moon

The Earth reminded us of a Christmas tree ornament hanging in the blackness of space. As we got farther and farther away it diminished in size. Finally it shrank to the size of a marble, the most beautiful marble you can imagine.

—James Irwin (1930–1991), Apollo 15 astronaut

Points to Ponder *(continued)*
… a small … planet … of a minor star … at the edge of an inconsiderable galaxy in the immeasurable distances of space … To see the earth as it truly is, small and blue in that eternal silence where it floats, is to see riders on the earth together, brothers on that bright loveliness in the eternal cold—brothers who know now they are truly brothers.

—Archibald MacLeish, American poet (1892–1982)
Procedure

Procedure

When Working With Teachers

Share the Points to Ponder quotes to elicit a discussion as to whether understanding science is antithetical to or provides an enrichment of an aesthetic and even poetic sense of wonder and appreciation for nature. Teachers can be encouraged to view any of the episodes from Carl Sagan's 1980 *Cosmos* series to see a paragon of an engaging and engaged scientist-teacher-poet. Also, the Annenberg Foundation's Private Universe and Harvard-Smithsonian Center for Astrophysics' MOSART professional development resources can be used to activate teachers' interest in the role of misconceptions in teaching and learning science (see Internet Connections for links to all three sources).

1. *Engage:* Use one or more of the following approaches to introduce the mini unit (i.e., depending on the amount of instructional time allocated for this phase):

   a. *Sentimental Songs and Scientific Slides:* Given its cyclical (i.e., approximately 29.5-day cycle of phases) ability to bring light into the darkest, most foreboding nights and its role in agricultural calendars and ocean's tides, the mysterious and seemingly changing Moon has inspired folklore, religious

stories, pseudoscientific speculations, and scientific study since the earliest days of architecture and art, written literature and music (e.g., Beethoven's 1801 "Moonlight Sonata" and Debussy's 1905 "Clair de Lune"), and natural philosophy and scientific treatises. Over the past 60 years, the Moon has been featured in many popular audio recordings, such as Frank Sinatra's "Fly Me to the Moon" (played during the Apollo 10 mission), Henry Mancini's "Moon River," Van Morrison's "Moondance," Cat Stevens' "Moon Shadow," Pink Floyd's "The Dark Side of the Moon," and Enya's "Shepherd Moons" (instrumental). Most of these songs associate the Moon with romance, madness, or other powerful human emotions.

A variety of videos posted on YouTube combine classic or new-age instrumental songs with striking Moon images (e.g., Beethoven, Debussy, or Enya). One of these types of videos can be compared and contrasted with Vangelis's "Albedo 0.39" (1976 title track), a song that presents an array of astronomical facts and figures about Earth. Different versions present this space age–sounding song against a backdrop of NASA images and equations related to the Earth-Moon system (see Internet Connections; note that Enya's *Moon* documentary "gives away" too much information for it to be used until the Explain phase). Color posters and 3-D models of Earth and the Moon provide a useful complement to elicit student prior conceptions about the relationship between Earth and the Moon, in particular the idea of the relative scale of the Earth-Moon system.

b. *Looney Tunes and Moon Humor:* Analysis of science cartoons and songs that feature the Moon and/or its relationship to Earth offer a lighthearted introduction to the topic (see Activity #12, Internet Connections: Sources of Science Cartoons). Similarly, Moon jokes can get learners thinking about the Moon in a different light (e.g., How can you see sunshine at night? and What name was given to the first satellite to orbit the Earth?).

c. *Man on the Moon in Movies:* Segments from science fiction and science fact films can be used to show how yesterday's dreams can become today's realities. See the Internet Connections, such as JFK Presidential Library and NASA sites and popular movies and TV series (e.g., *Apollo 13* and *From the Earth to the Moon*) to help set a context where the actual sizes and distance between the Earth and the Moon takes on a literal life-and-death relevance. *NSTA Reports* Blick on Flicks column (see Internet Connections) provides examples of how to separate fact from fiction in science fiction films.

d. *Moon Myths, Misconceptions, and Modern Lunar Lunacy?* Debunked Moon myths related to werewolves and "lunatics" notwithstanding, a variety of misconceptions are prevalent in modern culture. Teachers can develop a pre-instructional diagnostic assessment (e.g., statements with a scale of Disagree, Unsure, and Agree and/or multiple-choice questions) using misconceptions cited at the Internet Connections: Astronomy Diagnostic Test, Bad Astronomy, Heavenly Errors, and Harvard-Smithsonian MOSART websites. The objective is to assess learners' prior knowledge base (i.e., a mix of misconceptions, conceptual holes, and valid, foundational, conceptual precursors) for this mini unit or a broader unit on our solar system or astronomy in general. If desired, this assessment could be used as a pre- and postunit measure of the effectiveness of this mini-unit and/or to spark questions about the next one.

Focus Question 1: Do textbook illustrations (or commercial posters and models) accurately depict the relative sizes of or distances between objects (i.e., Earth, the Moon, planets, and the Sun) within our solar system? If not, why would they misrepresent or lie to us about scale?

Note: The answer to this question is to be developed via the 5E sequence; do not give the answer to the learners at this point. Instead, elicit their prior conceptions of the Earth-Moon scale and establish an atmosphere of FUNdaMENTAL science by "engaging" the learners with the following hands-on exploration (HOE). Hands-on activity does not necessarily

translate into efficient, effective learning. But HOEs that are also "minds-on" analogically turn over the fields of learners' minds to uproot weeds (misconceptions) and ready the soil for new seeds (scientific concepts). More directive, teacher-delivered explanations can and should wait until the third phase of the 5E Teaching Cycle, at which time learners are cognitively and emotionally ready to "hear," process, and integrate new "just in time" scientific ideas and terminology.

Sports Balls—Solar Science Scale Simulation: Hold up a "black garbage bag of science" and ask the learners what they think it might contain that would be useful for addressing the focus question. Open the bag and randomly distribute a variety of regulation-size athletic balls (i.e., one of each type listed in Table 13.1 in the Explain phase). Briefly discuss the idea of a scale model and ask the learners to hold the "sports spheres" over their heads so everyone can see them. Ask everyone to select and write down the two spheres that they think best represent the relative sizes of Earth and the Moon. Poll the class and tally, display, and discuss the results for the various possible combinations. Note the variety of "answers." Ask the class:

(1) Do scientists determine the validity of various claims by voting?
(2) If not, what numerical data would you need to determine the accuracy of your hypothesis, and where might we obtain such data?
(3) When we look at 2-D visual images or 3-D models of Earth and the Moon, do we perceive relative size in terms of linear (diameter), squared (surface area), or cubic (volume) measurements? Why? If desired, the PhET interactive simulation Estimation allows learners to make and check relative size estimations in one, two, and three dimensions (see Internet Connections).

Optional At-Home Experiment: The Moon Glows, But Does It Grow? If this 5E unit is started at or near the time of a full Moon, ask learners to observe and compare the apparent relative sizes of a rising full

Moon when it is near the horizon and again when it's near its peak elevation. If they repeat the experiment the next night using a 50 cm long dowel rod with a washer (with a 5 mm hole) taped down from the far end of the stick to "size up" the Moon, they will find that the Moon does not actually increase in size as it rises (as it appears to do to the unaided eye). Alternatively, students may curl their hand into a tunnel and observe the rising Moon with and without this tube effect, or they can take time-lapse photos of the Moon and measure its diameter. In each case, the optical illusion of objects appearing to be different sizes depending on the relative size and perceived nearness of the background is uncovered. This real world, discrepant-event experience, along with other perceptual illusions of Earth-bound observers, can lead to faulty predictions about astronomical relationships. Also, learners can use the internet or reference books to look up the actual equatorial diameters (i.e., most celestial bodies are not perfect spheres) for the Earth and the Moon for the next lesson. Later in the unit, consider showing the Smithsonian YouTube video clips on research that suggests the Moon has actually shrunk over time! Also, learners can begin a monthly log of the shape of the Moon that would extend into a subsequent unit where Moon phases are discussed.

2. *Explore:* Form small, cooperative groups, and give calculators and metric measuring tape or string and metersticks to every team so they can devise and test procedures for addressing the following questions:

   Focus Question 2: Given the actual equatorial diameters for the Earth and the Moon (12,756 km and 3,476 km, respectively), how can we determine which combination of two sports balls best represents the relative scale sizes of these two objects?

   a. In terms of *diameter*, how many times larger is Earth than the Moon? Devise a variety of nondestructive means of determining the diameter of the various sports balls to find the "best-matched" pair. *Note:* Different teams can be assigned to measure different balls.
   b. In terms of *volume* how many times larger is Earth than the Moon? Devise a variety of means of determining (or estimating) the volume of the various sports balls.

Note: Learning how to use the metric system and mathematical formulas is accomplished more efficiently when there is a real-world context or need for using the skills where they help answer a question or solve a problem of interest. Scientific investigations provide a "natural, need to know" context for developing such skills. If desired, the formulas that are given in the Answers section below could themselves become the focus of empirical investigations. For example, learners can discover π by measuring the circumferences and diameters of various objects and looking for a unifying pattern or experimentally check the formula for the volume of a sphere by using a water-filled balloon.

Optional History-Geography-Geometry-Science Story and/or Hands-on Exploration: As an alternative or supplemental aid to "giving" learners Earth's circumference (or diameter), consider sharing or having students experimentally replicate the story of Eratosthenes, the ancient Greek mathematician, geographer, astronomer, poet, and librarian. Eratosthenes was born in 276 BC in the Greek colony of Cyrene (now Shahhat in current-day Libya), appointed to run the Great Library of Alexandria in 240 BC, and died in 194 BC. After Eratosthenes read that at noon on June 21 (i.e., summer solstice or day with the maximum number of hours of daylight) the Sun was directly overhead and vertical sticks cast no shadow in Syene (near the Tropic of Cancer in modern-day Aswan, Egypt), he wondered if the results would be the same in Alexandria, which lay approximately 770 km (in modern metric units) north of Syene. In fact, vertical sticks were observed to cast shadows in Alexandria on this same day at the same time. Assuming that the Sun was so far away that its light hits the Earth at essentially the same time and angle, he reasoned that if the Earth was curved, the apparent discrepancy of no shadows versus shadows at the same time in different places could be resolved. With these assumptions, he calculated what curvature of the Earth could account for the length of the observed shadows and computed the spherical Earth's circumference to within 1.2% of the modern value.

This wonderful history and nature of science story, a detailed student worksheet, an internet-based project, and a teacher's guide are available from the Noon Day Project: Measuring the Earth's Circumference website (below). This site includes a 6:33-minute video clip from the classic 1980 PBS series *Cosmos* (from Episode 1, "The Shores of the Cosmic Ocean"), where Carl Sagan uses a model to demonstrate the logic of Eratosthenes's work. This discovery is also described in chapter 18 of *The Story of Science: Aristotle Leads the Way* by J. Hakim (2004). If desired, this story also provides an opportunity to contrast science and pseudoscience (see also the Internet Connections: Flat Earth Society and *In Search of the Edge: An Inquiry Into the Shape of the Earth* DVD; *More Brain-Powered Science*, Activity #22, Sample Test #1).

3. *Explain:* Pull the work groups back together to share their results. The following table—compiled from Microsoft's *Encarta CD-ROM Encyclopedia* (1994) data and derived calculations—can be compared to the class's empirical data and actual measurements of the relative diameters of Earth and Moon images from printed sources to see if the 3.67:1 ratio of Earth-to-Moon diameters is accurately depicted. As an approximation, project an image of a 10 cm circle (Earth) that has four 2.5 cm circles (or moons) lined up along the larger circle's horizontal diameter. Also, consider showing the fact-based Moon documentary that uses Enya's music as a backdrop (see Internet Connections: Enya's Music for 2010). The Explain phase is the "right time" for teachers to formally introduce scientific terminology that pulls together and extends the activities used in the Engage and Explore phases. Textbook readings, multimedia presentations, interactive lectures with active "note-making," concept mapping, and an array of other instructional strategies (see Appendix A) are appropriate during this phase. Well-placed and well-designed "direct teaching" can provide "just in time" cognitive scaffolding to support students' minds-on, cognitively active learning.

   Focus Question 3: If a picture is worth a thousand words, why is it important for textbook images such as the Earth-Moon system to be drawn to scale? Is it always possible to draw astronomical images to scale in textbooks?

Table 13.1. Sports Balls and Their Dimensions

Sports Ball	Circumference	Diameter	Radius	Volume
Basketball	76.0 cm	24.20 cm	12.10 cm	7,413.0 cm³
Soccer ball	71.0 cm	22.60 cm	11.30 cm	6,044.0 cm³
Volleyball	68.6 cm	21.80 cm	10.90 cm	5,425.0 cm³
Softball	30.5 cm	9.71 cm	4.85 cm	478.0 cm³
Baseball	23.0 cm	7.32 cm	3.66 cm	205.0 cm³
Tennis ball	20.9 cm	6.65 cm	3.33 cm	155.0 cm³
Racquetball	17.9 cm	5.70 cm	2.85 cm	97.0 cm³
Handball	15.0 cm	4.76 cm	2.38 cm	56.5 cm³
Golf ball	13.4 cm	4.27 cm	2.14 cm	40.8 cm³
Ping-Pong ball	12.0 cm	3.81 cm	1.91 cm	29.0 cm³

Table 13.2: Earth-Moon Dimensions and Sports Balls Scale Equivalents

Diameter of Larger Object	Diameter of Smaller Object	Ratio of Larger Object to Smaller Object
Earth 12,756 km	Moon 3,476 km	3.67:1
Basketball 24.2 cm	Tennis ball 6.65 cm	3.64:1
Volleyball 21.8 cm	Racquetball 5.7 cm	3.82:1

4. *Elaborate:* To *formatively assess* whether learners understand the basic idea of relative sizes (i.e., scale) and to return to and extend the original Focus Question #1, ask the following questions:

Focus Question 4a: Whether or not textbook drawings accurately reflect the relative sizes of Earth and its Moon, do you think they accurately represent the *relative distances* between Earth, its Moon, and other objects within our solar system?

Ask for two volunteers to assist you with the following visual analogue model:

Basketball–Tennis Ball Lunacy: A Lunar Orbiting Distance Participatory Demonstration: Using the basketball (Earth) and tennis ball (Moon) scale models agreed on in the Explain phase, ask the two volunteers to hold the balls so they are nearly touching. Ask the class if this represents the proper, scaled distance between the two. Ask the volunteer holding the basketball (Earth) to slowly move away from the tennis ball (Moon) (approximately one basketball diameter per step) and have the rest of the learners raise their hands when they think the proper orbiting distance has been reached. *Note:* Most people will raise their hands long before the approximately 30 basketball measures that should separate the two balls on this scale. As you reach the limit of the length or width of your classroom, rather than giving them this answer, have learners calculate the answer to the following questions.

Focus Question 4b: Given an *average* (i.e., the orbit is not perfectly circular) Earth to Moon distance of 384,400 km, how far apart should the two balls be to reflect reality on this scale? *Note:* To achieve the 30-basketball separation distance, you may need to use a diagonal in the classroom (or take the learners into the hallway).

Focus Question 4c: Could a textbook of typical dimensions correctly show *both* the relative size and relative distance of the Earth-Moon system within the confines of a two-page layout? Could other planets in our solar system be shown on this same scale? Explain your answers with appropriate mathematics. How can scale models overcome textbook limitations?

Optional Mathematical Comparisons of the Earth and Moon: Consider having students explore other mathematical comparisons such as their relative mass, density, and gravitational force and how these ratios compare to their relative sizes. Students could also consider how the relative sizes and orbital distance compare to other planet-moon combinations in our solar system. *Note:* The Moon is both relatively larger and farther away from Earth than is true for other planet-moon combinations. Such considerations can lead to an exploration of alternative theories of the origin of our Moon. Further afield, learners can begin to explore the unique Sun-Earth-Moon systems relationships that make Earth life possible. Consider the analogical links between Earth (the only planet known to contain both a prominent hydrosphere and a rich, diverse biosphere), the story of Goldilocks, and the phrase "Spaceship Earth." What astronomical, geological, and chemical factors make our planet "just right" to support life? The Points to Ponder quotes are useful in this context.

Optional Mathematical Excursions: Exponential Notation and Analogies for "Millions": Both the diameter of the Earth and its Moon are measured in millions (10^6) of meters. Practice with exponential

notation and visual representations and analogies can make the concept of orders of magnitude less abstract. The children's book *How Much Is a Million* by David M. Schwartz and Steve Kellogg (1985) suggests a variety of ways to conceptualize a million, such as that a fish bowl big enough to hold a million goldfish would be large enough to hold a 60 ft. whale. Challenge students to come up with their own representations for a million. A simple model that can be displayed on the classroom walls is dots (as many as 4,000–5,000) on standard 11 in. × 8.5 in. stationery (for a total of 250; 200 pages will represent 1 million). Additionally, the *Powers of Ten* CD-ROM, videotape, and books present a visualizing stunning look at "the relative size of things in the universe" from the 10^{25} to 10^{-16} meter scale known to modern science. See also the websites listed in Extension #2 (on p. 248), Activity #12 in this book, and Activity #17 in *More Brain-Powered Science* (describes how a cubic meter contains 10^6 cm^3 or 10^9 mm^3) for other models of powers of 10 magnitudes and real-world applications.

5. *Evaluate:* Though well-designed conventional paper-and-pencil tests can be a critical component of summative assessments (e.g., pre- and posttest measures of gains in scores when compared to diagnostic tests), other alternative assessments are also valuable. The point about the relative emptiness of space can be brought home to students by having them complete one or more of the following scale-model-type activities:

  a. *Daytime Astronomy With Balls and Sticks:* Design and construct a handheld scale model of the Earth-Moon system. For example, the following model works quite well: Use a 122 cm long and 2.3 cm diameter dowel rod with two holes separated by 117 cm. A Ping-Pong ball (Earth; d = 3.81 cm) is glued to the top of one 8 cm stick and a 1 cm wooden bead (Moon) to the top of another. This model can be used to demonstrate (outdoors on a clear, sunny day) differential lighting on various portions of the Earth (i.e., day versus nighttime areas of the globe), the phases of the Moon, and so on. See Internet Connections: NSTA's Astronomy With a Stick/Day Into Night for other daytime astronomy activities.

Table 13.3. "Far-Out" Interplanetary Comparisons

Body	Equatorial Diameter	Body	Mean Distance From Sun
Mercury	4,800 km	Mercury	57,900,000 km
Pluto*	6,000 km	Venus	108,000,000 km
Mars	6,787 km	Earth	149,600,000 km (1 AU)
Venus	12,104 km	Mars	229,000,000 km
Earth	12,756 km	Jupiter	779,000,000 km
Neptune	49,500 km	Saturn	1,427,000,000 km
Uranus	51,800 km	Uranus	2,871,000,000 km
Saturn	120,000 km	Neptune	4,496,000,000 km
Jupiter	142,800 km	Pluto*	5,913,000,000 km
Sun	1,393,900 km		

* Technically no longer considered a planet

b. *Solar System Soccer Scale:* Have learners obtain data from Internet sites (or Table 13.3) to see if they can design a scale model of our solar system that reflects both relative sizes and interplanetary distances that will "fit" on a 100 m long soccer field (or roll of toilet paper or other long, standard-size object). If so, can all the planets be seen with the naked eye from any location in the solar system (soccer field)?

Note: Beyond our solar system, the interplanetary and interstellar distances become so immense that they are measured in light years (i.e., the distance that light travels in one year based on a constant velocity of 300,000 km/sec.). For comparison, consider that it takes light approximately 8 minutes to travel from the Sun to Earth, and the Sun's light reflected off the Moon reaches Earth in 1.3 light seconds. When astronomers look at distant stars and galaxies from Earth, they are really looking back into the history of what once was, not at the stars or galaxies in their present conditions or locations (i.e., they may no longer even exist).

c. Curious Calculations: A Homo Sapien–Solar System Scale

   (1) If the Sun was represented as a 6 ft. diameter sphere (a little more than an average adult male's height), what body part would represent Jupiter, the largest planet, and Earth? What's wrong with this model?

(2) If the Sun was a sphere whose diameter equaled the size of one large human's foot (approximately 12 in.), what body part would the Earth be equivalent to in size, and how far away would it need to be placed?

d. Silly Science Songs and Solar System Science Surveys

(1) Rewrite the lyrics to popular Moon-themed songs to directly confront and correct the common misconceptions and/or teach correct scientific facts that were explored in this 5E Teaching Cycle. Public performances are optional but can be highly engaging.

(2) Develop, administer, and analyze a schoolwide survey of students, faculty, and/or parents to see how common astronomical misconceptions are.

Debriefing

Appendix C compares the *Benchmarks for Science Literacy* (AAAS 1993) Common Themes, the *National Science Education Standards* (NRC 1996) Unifying Concepts and Processes, and the draft *Framework for Science Education*'s Cross-Cutting Scientific Concepts (NRC 2010a). The table reveals that over the past 17 years, major scientific organizations have consistently agreed that "big ideas" such as systems, models, and scale that are featured in this 5E should be emphasized within and across all of grades K–12. Although not explicitly addressed in this unit, the idea of constancy and change is in fact a subtext for it and would naturally be featured in subsequent astronomy-related units. The ancient Greek idea of *cosmos* (order) amidst *chaos* (disorder) evolved as natural philosopher-astronomers studied the movements of planets and stars and remains a guiding principle for scientific research to this day. This mini unit provides an opportunity to model the "forest for the trees" perspective with and for teachers and/or their students. It also demonstrates how science is FUNdaMENTAL and involves both "fun work" and "hard play" and how analogical models can use the "familiar" (e.g., sports balls) to understand the "strange" (Earth-Moon scale).

This mini unit also invites learners to become active, critical thinkers and reviewers (versus passive, mindless absorbers and consumers) of the "truths" represented in textbook visuals (i.e., a picture is worth a thousand words). Regrettably, the "aura of authenticity and authority conveyed by very elaborate and colorful illustrations, but with erroneous content, are especially invidious, because they will reinforce flawed or incomplete prior knowledge" (Mathewson 1999, p. 43). The relatively vast amount of "nothing" (truly empty space) that dominates reality at the scale of the submicroscopic inner worlds of atoms and the interplanetary and intergalactic worlds of outer space cannot be accurately portrayed within the dimensions of textbooks. By necessity, textbook images grossly misrepresent this highly counterintuitive fact. But they also typically fail to either post a warning on the images or suggest models, computer animations, multimedia simulations, or participatory demonstrations that correct their visual misrepresentations.

Conceptual change requires learners to become dissatisfied with their current conceptions (due to conceptual holes and/or anomalous discrepancies) and see the alternative scientific model as more intelligible, plausible, and fruitful (Posner et al. 1982). "Questioning the answers" creates a classroom community based on a "dialogue of discovery" rather than a "monologue of conclusions." Effective science instruction requires inspiration rather than indoctrination. Given the interrelated nature of scientific theories and explanations, subsequent units will need to return to reinforce and extend the idea of "big empty spaces" that are introduced in this mini unit.

Following the Evaluation phase, this 5E Teaching Cycle would subsequently serve as a lead-in to additional units that focused on concepts such as the phases of the Moon; reasons for the seasons; lunar eclipses; the law of universal gravitation; historic and current-day uses of telescopes; and the age, size, and origin of the universe. Analogous to the way that early telescopes "opened the eyes and minds" of 17th-century Europeans (Panek 1998), astronomy units can help students "see" that the cosmic scales of space and time are truly mind-altering and awe-inspiring. Many astronauts and scientists have been influenced to more profoundly value "Spaceship Earth" based on the broader perspective their journeys of discovery have afforded them (see the Points to Ponder quotes). This ongoing (and much parodied) theme runs through all of Carl Sagan's narrations in his classic 1980 *Cosmos* PBS series.

Extensions

The Internet Connections contain a wide variety of resources for expanding this 5E mini unit and developing other astronomy units that take learners beyond the immediate celestial "backyard" of our Moon. For example, consider the following activities:

1. *Multimedia Mayhem and Serious Science:* Varying degrees in the humor and science of larger-scale astronomical facts and figures can be found in multimedia such as (a) "The Galaxy Song" (from Monty Python's *The Meaning of Life* movie); (b) Enya's music for science documentaries; and (c) *Bill Nye the Science Guy*'s eight 26-min. episodes related to astronomy (which include "Comets and Meteors," "Earth's Seasons," "Gravity," "Moon," "Outer Space," "Planets," "Space Exploration," and "the Sun."). They are available for purchase from Disney Educational Productions for $29.99 per DVD (targeted for grades 3–8).

2. *Super-Sized Scales of Science:* "Mind-blowing" powers-of-10-type animations can be found at the Cosmic View, Microcosm, Powers of Ten, Secret Worlds, and The Known Universe internet sites (see Internet Connections). These animations can be revisited in multiple units during the school year.

3. *Moon Watchers and Star Gazers:* Students often mistakenly assume that all the great scientific discoveries occurred in the modern era. Teachers and students can explore the history of our understanding of the Earth-Moon relationship through the internet and books such as Joy Hakim's *The Story of Science: Aristotle Leads the Way* (2004; see especially chapters 3, 9, 13, and 18) and *The Story of Science: Newton at the Center* (2005; see especially chapters 1–11). Also see the previously mentioned *Cosmos* TV series and book by Carl Sagan. Historical events to explore include those listed below:

   • Approximately 3000 BC: Babylonians develop a 12-month, 29.5-day lunar calendar (which unfortunately doesn't keep pace with the agriculturally important, seasonally linked solar cycle that was later developed by the Egyptians).

248

- 500 BC: Pythagoreans teach that the Earth is a 3D sphere rather than a 2D flat disk and that the "kosmos" (i.e., cosmos) is ordered like music and mathematics.
- 330 BC: Pytheas, a Greek geographer, notes a connection between the Moon and the tides.
- 270 BC: Aristarchus of Samos proposes a heliocentric model and estimates the Earth-Sun distance.
- 240 BC: Eratosthenes of Cyrene correctly calculates the Earth's circumference.
- 130 BC: Hipparchus of Nicea correctly calculates the distance to and size of the Moon.
- 1580: Tycho Brahe, the last great naked-eye astronomer and practicing astrologer, relies on Ptolemy's geocentric model and rejects Copernicus's heliocentric model.
- 1609: Galileo uses his hand-built telescope to see craters and mountains on the Moon, demonstrating that heavenly spheres are not necessarily "perfect" or different in nature from Earth. He was a strong advocate for the heliocentric model.

Internet Connections

- American Astronomical Society, Education Services: *http://aas.org/education*

- Annenberg Foundation: *A Private Universe* (1987; 20 min. professional development video-on-demand that helped launch research on role of misconceptions in learning science): *www.learner.org/resources/series28.html*

- Association for Astronomy Education: Resources (including astronomy diagnostic tests): *www.cis.rit.edu/~jnspci/AAE/resources.shtml*

- Astronomical Society of the Pacific, Education: *www.astrosociety.org/education.html*:

    Free and for-sale resources (e.g., *Universe at Your Fingertips* activity book and Moon maps): *www.astrosociety.org/education/astro/astropubs/universetoc.html*

- Astronomy Café (books and more): *www.astronomycafe.net*

- Astronomy Course for Middle/High School Students: *http:// darkskyinstitute.org*
- Astronomy Diagnostic Test (21-item free download): *http://solar. physics.montana.edu/aae/adt*
- Astronomy Picture of the Day: *http://antwrp.gsfc.nasa.gov/apod/ astropix.html*
- Bad Astronomy (author Phil Plait's site on misconceptions and more): *www.badastronomy.com/index.html*
- BBC Science and Nature: Space (video clips, simulations, games, and more): *www.bbc.co.uk/science/space*
- Beethoven's "Moonlight Sonata" and Moon (photo) Montage: *www.youtube.com/watch?v=c1tkBNF0ghU* (2:48; piano)
- CERES Project (NASA-funded Center for Educational Resources for astronomy education): *http://btc.montana.edu/ceres*
- Challenger Center for Space Science Education (search: Lessons): *www.challenger.org*
- Cheryl Robertson's Whole Moon Page (Moon in ancient and modern history, religion, arts, legends and tales, physical characteristics, popular songs lyrics, etc.): *www.moonlightsys.com/ themoon/index.html*
- Concord Consortium: Molecular Workbench (free downloadable simulations):

    Our Solar System (Middle School): *www.concord.org/activities/ our-solar-system*

- *Cosmic View: The Universe in 40 Steps* (online version of 1957 book by Kees Boeke): *http://nedwww.ipac.caltech.edu/level5/Boeke/ frames.html*
- *Cosmos* (Carl Sagan's 1980 PBS series; see episode 1, "The Shores of the Cosmic Ocean," for the segment on Eratosthenes; see also Noon Day Project below): *www.hulu.com/cosmos* or search Google Videos
- Debussy's "Clair De Lune":

    www.youtube.com/watch?v=CvFH_6DNRCY&feature=related (4:59 min. with static Moon) *www.youtube.com/watch?v=ZfSV_k3MhCw* (4:56 min. with multiple images of Moon)

- Disney Educational Productions: *Bill Nye the Science Guy*: *http://dep.disney.go.com*

- Enya's "Shepherd Moons" (3:41 min. title track):

    www.youtube.com/watch?v=1iRpa-hi1N4 (with images of the Moon)

    www.youtube.com/watch?v=SM9UWnzzAoc&feature=related (with mythical images)

- Enya's Music for 2010 Moon documentary (5:39 min.; images, statistics, and astronaut audio clips): *www.youtube.com/watch?v=_tsIcIDI3mo*

- Enya's Music for 2010 Travel in the Universe documentary (10:00 min.; images and printed text): *www.youtube.com/watch?v=0v26UD BqUrE&feature=related*

- Exploring Planets in the Classroom (various activities, including the Moon): *www.spacegrant.hawaii.edu/class_acts/index.html*

- Flat Earth Society:

    http://theflatearthsociety.org/cms

    http://en.wikipedia.org/wiki/Flat_Earth_Society

    www.alaska.net/~clund/e_djublonskopf/FlatHome.htm

- Fourmilab Switzerland (Astronomy and Space: Earth and Moon Viewer, Solar System Live, and Your Sky interactive programs): *www.fourmilab.ch*

- Google Earth: *http://earth.google.com* (zoom down to your school with this free download)

- Harvard-Smithsonian Center for Astrophysics:

    How Big Is Our Universe?: *www.cfa.harvard.edu/seuforum/howfar/howfar.html*

    MOSART: Misconception Oriented Standards-based Assessment Resource for Teachers: *www.cfa.harvard.edu/smgphp/mosart/about_mosart.html*

- Heavenly Errors (author Neil F. Comins's site on Misconceptions: Solar System: Moons): *www.physics.umaine.edu/ncomins*
- *In Search of the Edge: An Inquiry Into the Shape of the Earth* (media literacy "documentary"): *www.bullfrogfilms.com/catalog/search.html*
- Jet Propulsion Lab, Solar System Dynamics Group (physical characteristics and orbits): *http://ssd.jpl.nasa.gov*
- John F. Kennedy Presidential Library and Museum: Space Program: *www.jfklibrary.org/Historical+Resources/JFK+in+History/Space+Program.htm*
- *Microcosm*, CERN's Interactive Science Centre: Powers of Ten: steps through 10^{-15}–10^{26} m views: *http://microcosm.web.cern.ch/Microcosm/P10/english/welcome.html*
- Monty Python's "The Galaxy Song" (from *The Meaning of Life* movie):

  Lyrics: *www.lyricsdepot.com/monty-python/galaxy-song.html*

  Wikipedia: Analyzing the scientific validity: *http://en.wikipedia.org/wiki/Galaxy_Song*

  YouTube (2:44 min. original video clip and slide show with printed lyrics):

  www.youtube.com/watch?v=buqtdpuZxvk (original movie clip)

  www.youtube.com/watch?v=J5LMbcoTFFE&feature=related (song and lyrics)

- NASA Apollo 40th Anniversary (varied resources such as Discovering Earth PDF images, We Choose the Moon, and NASA Benefits You on Earth): *www.nasa.gov/mission_pages/apollo/40th/index.html*
- NASA Home (for Educators): *www.nasa.gov/audience/foreducators/index.html*
- NASA Space Science Education Resource Directory: *http://teachspacescience.org/cgi-bin/ssrtop.plex*
- NASA Space Store and Gift Shop (posters and more): *www.thespacestore.com*

- NASA Video Clip of Neil Armstrong stepping onto lunar surface (note: slow connection): *www.hq.nasa.gov/office/pao/History/alsj/a11/a11v_1092338.mpg*

- ~~Nine~~ 8 *Planets: A Multimedia Tour of Solar System: http://nineplanets. org*

- Noon Day Project: Measuring the Earth's Circumference (includes 6:33 min. video clip of Carl Sagan from *Cosmos* on Eratosthenes): *www.ciese.org/curriculum/noonday*

- Nova Celestia (free astronomical illustrations and space art): *www.novacelestia.com*

- NOVA Online: Runaway Universe: *www.pbs.org/wgbh/nova/universe*

    To The Moon (broadcast 6/13/99): *www.pbs.org/wgbh/nova/tothemoon*

- NSTA's Astronomy With a Stick and Day Into Night (interactive resources): *www.nsta.org/publications/interactive/aws-din*

- *NSTA Reports* Blick on Flicks column: *www.nsta.org/publications/blickonflicks.aspx*

- PhET Interactive Simulations: Estimation (explore relative size in 1, 2, and 3 dimensions): *http://phet.colorado.edu/en/simulation/estimation*

- Physlink.com: Physics and Astronomy Online (education, humor, and more): *www.physlink.com/Index.cfm*

- *Powers of 10* (1977 film journey from 10^{25}–10^{-18}m): *http://powersof10.com*

- Sciencemall-usa (posters): *www.sciencemall-usa.com*

- Science Posters Plus: *www.super-science-fair-projects.com/science-posters-plus.html*

- Sea and Sky: Astronomical Art: *www.seasky.org/links/skylink08.html*

- *Secret Worlds, the Universe Within*: free, interactive Java "powers of 10" type video: *http://micro.magnet.fsu.edu/primer/java/scienceopticsu/powersof10*

- Smithsonian National Air and Space Museum (varied resources): *www.nasm.si.edu*

- Snopes.com: One Small Misstep (Apollo 11 story): *www.snopes. com/quotes/onesmall.asp*

- Sources of Science Cartoons: See listing in Activity #12.

- Space Telescope Science Institute: *www.stsci.edu/portal*

  A variety of resources, such as *Comparing Earth and Its Planetary Neighbors*: *www.stsci.edu/exined/Comparing.html*

- Starry Night (simulation software for purchase): *www.starrynight. com*

- *The Known Universe* (American Museum of Natural History "powers of 10" type film travels from the Himalayan Mountains out 13.7 billion light years and back to Earth in 6:31 min.):

  http://apod.nasa.gov/apod/ap100120.html

  www.youtube.com/watch?v=17jymDn0W6U

- University of Iowa Physics and Astronomy Lecture Demonstrations (see Scale Model Demos of the Solar System and Scale of the Earth/Sun/Moon System): *http://faraday.physics. uiowa.edu*

- Vangelis—"Albedo 0.39" (4:20 min. title track with Earth statistic–type lyrics overlaid on NASA images of Earth-Moon system):

  www.youtube.com/watch?v=rL1oU6fH25w&feature=related

  See also:

  www.youtube.com/watch?v=kmR8kWotPWg (5:30 with images and math)

  www.youtube.com/watch?v=2Q4lR0cY0fc&feature=related (lyrics on one images)

- Views of the Solar System (animations, images, information and posters): *www.solarviews.com*

- We Choose the Moon (relive Apollo 11 mission in 11 stages with extended animation, archival photos, video, and mission audio type): *www.wechoosethemoon.org*

- Wikipedia: *http://en.wikipedia.org/wiki*. Search topics: *Apollo 13* (1995 film), *Cosmic Voyage* (1996 IMAX film that depicts 42 orders of magnitude; similar to *Powers of 10*), *From the Earth to the Moon* (Jules Verne's science fiction story, 1958 movie, and 1998, 12-part HBO miniseries on the *Apollo* missions), Goldilocks, orders of magnitude (length), parts-per notation, *Powers of 10* (film), and Spaceship Earth
- Windows to the Universe (animations, images, and activities): *www.windows2universe.org*
- YouTube: (Smithsonian Science clips):

    Meet Our Scientist: Tom Watters: *www.youtube.com/ watch?v=7avaIqEiygc*

    Hints of a Shrinking Moon: *www.youtube.com/ watch?v=oE6uxsLOzMI&NR=1*

Answers to Questions in Procedure, steps #1–#5

1. *Engage:*

    a. Scientific conclusions are not reached by a democratic majority rule voting process. Nor is there the equivalent of a single, representative elected body of "science senators" who rule on scientific validity (although the editorial boards of professional journals, funding agencies, and federal regulations do act as selection filters for public and privately funded research). Tentative, provisionally accepted scientific ideas are based on *empirical evidence* (e.g., this commonly includes numerical data), *logical argument* (e.g., that typically relies on an interwoven tapestry of related theories), and *skeptical review* (which leaves consensus views open to further debate and modification.

    b. At the end of the Engage phase, data on the size of the Earth and Moon can be given to the learners from an authoritative source (though an optional recreation of Eratosthenes's experiment is described in the Explore phase).

c. When looked at from a distance, 3-D spheres look more like flat, 2-D disks, and people tend to perceive differences in sizes as being related to the diameters rather than surface areas or volumes, which are much harder to estimate with accuracy. *Note:* Textbook visuals of the Earth-Moon system are typically not drawn to scale. Also, even if the relative sizes of celestial bodies are drawn to scale (a rarity), relative interplanetary distances of other objects within our solar system are not and cannot be drawn within the confines of a textbook without the use of a huge, costly foldout section (or running the relative scale across the bottom of many pages).

2. *Explore:*

   a. Possible approaches for determining the balls' diameters include the following:

      (1) Use a string and metric stick or tape measure to measure the circumference and use this equation:

      Circumference (or C) $= \pi \times$ diameter (or d) to calculate d

      (2) Place an ink mark on the ball and roll it along a metric stick to measure C directly.

      (3) Place the ball on a metric stick, sandwich the ball between two vertically placed and perfectly perpendicular books that rest on the ruler, and measure the diameter.

      (4) Use published reference materials to find regulation dimensions of the balls (see data provided below). The ratio of Earth-to-Moon diameters $= 12{,}756/3{,}476 = 3.67/1$.

   b. The ratio of Earth-to-Moon volumes $= (3.67)^3/1^3 = 49.4/1$. Possible approaches for measuring the balls' volumes include the following:

      (1) For small, denser-than-water golf balls, use simple water displacement in a large graduated cylinder or beaker.

      (2) Similarly, the Ping-Pong ball and other "lighter-than-water" balls can be submerged with heavy weights of known volume.

      (3) Larger "lighter-than-water" balls can be submerged in a filled bucket of water equipped with an overflow lip and catch basin.

(4) Hollow, relatively fixed-shape balls (Ping-Pong balls, racquetballs, handballs, and tennis balls) can be sliced in half or have holes drilled in them to fill with water. *Note:* The measurement of internal volume will be somewhat less than external volume depending on the thickness of the ball's "skin" and would also result in a low estimate for inflated balls such as volleyballs, soccer balls, and basketballs.

(5) Use the formulas: Radius = diameter/2 and Volume = $4/3 \pi (r^3)$.

3. *Explain:*

Focus Question 3: Human conceptions of reality are strongly influenced by our sense of sight, as evident in expressions such as "If I see it, I'll believe it" and "I see" (same as "I understand"). Limitations of human visual perception are a source of some of our most persistent misconceptions. The extreme range of orders of magnitude (or powers of 10) of the various levels of reality (including vast expanses of both inner and outer space) often make it necessary that textbooks and other visual media "lie to us." When this is necessary in textbooks, a "truth in advertising" written statement such as "NOT drawn to scale" should be affixed to each "incorrect" image, and analogical models and simulations should be suggested. From the scale of subatomic particles through cosmos, science textbooks grossly undersell the "far out" and truly awesome nature of reality as uncovered by scientific exploration.

4. *Elaborate:*

Focus Question 4b: 384,400 km Earth-Moon separation/12,756 km Earth diameter = 30/1 or 30 Earth (basketballs) separate the Earth (basketball) from the Moon (tennis ball). Focus Question 4c: Yes, assuming a 40 cm two-page layout, a 0.25 cm diameter Moon, a 1 cm Earth, and a 30 cm separation between the circles, both the relative size and distances of the Earth-Moon system could be displayed on the same scale across two adjacent pages. To the author's knowledge, no middle or high school science textbook illustration shows both the relative Earth-Moon

diameters and distance to the same scale. If one wants to represent the full orbit of the Moon around the Earth, the scale would need to be reduced to a 1 mm diameter Moon. Scale drawings of the orbits of other bodies in the solar system are not possible within the confines of textbook dimensions. Unfortunately, few, if any, textbooks tell students that they "have to" misrepresent scale given the fact that outer space is mainly "empty," with immense distances between celestial bodies (nor do they provide scaling exercises for students to get a more accurate sense of this). Accordingly, students walk away from instruction with an impoverished underestimation of the full wonder of the universe and our place in it.

5. *Evaluate:*

c. (1): If the Sun had 6 ft. diameter, a person's head would represent Jupiter and the iris of the eye would be Earth. Of course, the relative distances between the two planets would not be anywhere close to scale.

c. (2): Alternatively, if the Sun's diameter were 1 ft., Earth would be about the size of a human skin mole (1/8 in.) and would need to be placed 100 Sun diameters (or feet) away to show the proper orbiting distance.

Appendix A

ABCs of Minds-on Science Teaching (MOST) Instructional Strategies

Popular culture often equates teaching with telling and learning with listening. In addition to "translation errors" that occur between an intended message and a speaker and one or more listeners, teaching that relies predominately on one-way, verbal-only "transmission and passive reception" fails to acknowledge (1) the differences between information, knowledge, understanding, wisdom, and behavioral changes; (2) variations in learners' perception and processing of visual, auditory, and kinesthetic information; and (3) the *interactive* (i.e., student-phenomena, student-teacher, and student-student) nature of learning as an act of constructing personally meaningful and transferable knowledge that can be applied in creative ways and extended into new contexts.

Current research emphasizes learners' cognitively active roles in selecting, organizing, reconstructing, and integrating new information in light of the learners' prior knowledge. Our sensory buffers and working and short-term memories have limited capacities in terms of the "cognitive load" (i.e., relative quantity and complexity of content) that can be processed in a given time frame. However, as visual and auditory inputs appear to be initially processed as separate data streams, teaching that intentionally and intelligently integrates "sights and sounds" can enrich without overloading students' processing of essential (versus extraneous) information (see *Brain-Powered Science* and *More Brain-Powered Science* for additional activities that demonstrate these ideas and Mayer [2009a] for a synopsis of research on multimedia learning and instructional design principles). Additionally, it is known that "hands-on," laboratory-based activities (and/or those that include other bodily motions)—though potentially pedagogically powerful—do not necessarily automatically translate to "minds-on" learning (Clark, Clough, and Berg 2000; Clough 2002; Clough and Clark 1994a, 1994b; Colburn 2004; Doran et al. 2002; Hofstein and Lunetta 2003; Lunetta, Hofstein, and Clough 2007; NSTA 2007c; Singer, Hilton, and Schweingruber 2006; Tobias and Duffy 2009; Volkman and Abell 2003). Similarly, "hands-off" is not necessarily the antithesis of "minds-on" (see Table A below). As the adages "Variety is the spice of life" and "Different strokes for different folks" suggest, teachers can "reach and teach" diverse students by drawing on an extensive and continuously evolving "toolkit" of instructional strategies (rather than on a "monomaniacal, one size fits all" pedagogy).

Brainstorming a list of A to Z instructional strategies that can activate attention and catalyze cognitive processing and thereby facilitate active learning is a useful professional development exercise. In reflecting on their individual or collective lists, teachers may discover that the line between instructional strategies and curriculum-embedded diagnostic and formative assessments is a fuzzy one in that the best instructional activities typically provide feedback to both teachers and learners on the relative effectiveness of their efforts (Doran et al. 2002; Enger and Yager 2001; Liu 2010; Mintzes, Wandersee, and Novak 2000; NRC 2001b, 2001c, 2006). In any case, the MOST learning occurs when students come to class with anticipation and leave with regret rather than the reverse. Across a given instructional day, 5E unit, and academic-year course, teachers who creatively integrate multiple strategies and real-world connections and applications increase the probability that the cognitive and emotional needs of diverse learners will be met (CCSSO 2010; NBPTS 2003a, 2003b; NRC 2010b; PCAST 2010).

The importance of diverse instructional strategies as part of the science and art of science teaching was recognized more than 120 years ago:

> The search for instructional methods has often proceeded on the presumption that there is a definite patent process through which all students might be put and come out with results of maximum excellence; and hence pedagogical inquiry has largely concerned itself with the inquiry, "What is the best method?" rather than with the inquiry, "What are the special values of different methods, and what are their several advantageous applicabilities in the varied work of instruction." The past doctrine has been largely the doctrine of pedagogical uniformitarianism …
> (Chamberlin 1965/1890, p. 757)

But performance skills (Tauber and Sargent Mester 2007) and multiple methods notwithstanding, master teachers also know that "Good teaching cannot be reduced to technique; good teaching comes from the identity and integrity of the teacher (p. 10) … Good teachers … are able to weave a complex web of connections among themselves, their subjects, and their students so that students can learn to weave a world for themselves (p. 11)" (Palmer 2007).

Weaving new strategies into one's instruction is a process that requires time, practice, feedback, and collegial support to reach a level of professional competence and confidence that is comparable to strategies one has already mastered (DuFour and Eaker 1998; Loucks-Horsley et al. 1998; Mundry and Stiles 2009; Yager 2005). Similarly, students need support to "learn how to learn" in new ways that may initially feel awkward or uncomfortable. Effective teachers are not confined to or limited by their past successes and current competencies, but rather are always learning along with their students. Learning involves stretching the limits of one's comfort zone and working through implementation dips in performance. It is typically facilitated by a *scaffolded immersion* rather than a "sink or swim submersion." Journals, books (e.g., the *Brain-Powered Science* series),

internet resources, professional associations, and collaborations with colleagues are all valuable sources of professional development to counteract individual and institutional inertia and assist teachers in learning how to effectively incorporate new instructional strategies in pedagogically powerful ways. The effectiveness of implementation of any instructional strategy depends on the answers to these questions:

WHO are the teachers and students (e.g., their age, prior experiences, skills, and attitudes)?

WHAT are the relative pros and cons of the strategy (e.g., opportunity costs in terms of time and resources not available for other objectives)?

WHERE or in what kind of learning environment will the strategy be used (e.g., classroom, lab, on or off-school field site, museum, computer-based virtual environment, or at home)?

WHEN is the strategy used in a curriculum-instruction-assessment (CIA) sequence in a unit/course?

HOW frequently will the strategy be used, and how will its effectiveness be evaluated?

WHY is the strategy being used relative to the targeted curriculum objectives?

A→Z Instructional Strategies

Acronym, Acrostic, Activity centers, Advance organizer, Analogy, Animations, Anomalies, Anticipation guides (for reading), Apprenticeships, Audiotapes, and Authentic and Alternative Assessments

Biographies, Blogs, Books, Brainstorming, Brainteasers, Brochures, and Bulletin Boards

Calculator-based labs (CBL), Cartoons/comics, Case study, Charades, Cognitive coaching, Cognitive conflict (or dissonance), Collaborative and Cooperative learning, Collages, Computer-assisted instruction (CAI), Concept attainment model, Concept mapping, Conferencing, and Crossword puzzles

...ABCs of Minds-on Science Teaching (MOST) Instructional Strategies

Dance, Database Discoveries, Debate, Design projects, Diagnostic assessments, Dioramas, Discrepant demonstrations, Discussion, Displays, Distance learning, Drama, and Drawing

Event-based explorations, Examples, Exercises, Exit slips (or Entry cards), and Experiments

Field studies, Films, Fishbowl discussions, Flash cards, Flipcharts, Flowcharts, and Formative assessments

Games, GIS (geographic information systems) and GPS (global positioning system), Graffiti wall, Graphical analysis, Graphic organizers, Guest speakers, and Guided practice and reading

Hands-on explorations, Historical vignettes and skits, Homework, and Humor

Icebreakers, Interactive illustrations, Independent study/research, Inquiry-based investigations, Interactive video, Interest surveys, Internet-based inquiry, Internships, Interviewing, and Issue-based investigations (typically science-technology-society)

Jeopardy, Jigsaw (puzzles and cooperative learning), Jingles, Jokes, Journaling, and Just-in-time teaching

Kinesthetic activities and KWL charts (and KWL Plus with a graphic organizer as a summary)

Laboratory investigations, Learning centers, Learning cycles, Learning logs, Lively lectures, Letter writing, Likert scale opinion surveys (SD D U A DA), and Literature circles

Manipulatives, Maps, Mastery learning, Mentoring, Metaphors, Microcomputer-based laboratories (MBL), Microscopy, Mime, Mind mapping, Mnemonics, Mock news and trials, Models, Multimedia and/or Multimodality-based learning, Museum visits, and Music

Nature walks, Neurobics, Newscast and Newsmedia analyses, Note-taking, and Non-linguistic representations

Observation of phenomena, Observation logs, Oral histories, and Oral presentations

Panels, Peer teaching and learning, Performance assessment, Plays, Podcasting, Poetry, Polls, Portfolios, Posters, PowerPoint Presentations (teacher- and student-designed), Predict-observe-explain (POE), Problem- and Project-based learning (PBL), Probeware, and Puzzles

Quandaries, Questionnaires, and Questions (Bloom's taxonomy and student- and teacher-generated)

Rap music, Read-alouds, Recitation, Reciprocal teaching, Re-enactment/ Role-playing, Reflective journals, Reports, Research projects, Review, Riddles, and Rhymes

Science fairs, Science-writing heuristic (SWH), Semantic webbing, Service-learning, Shadowing, Show and tell, Simulations, Singing, Skits, Sociodrama, Socratic dialogue, Sponge activities, Spreadsheets, Storytelling, Structured controversies, Study groups, STS activities, Summarizing strategies, Surveys, Symposium, and Synectics

Technology-enabled instruction, Telecollaboration/Teleconferencing, Tests that teach, Think aloud, Think-Write (/or Draw)-Pair-Share, TOYS (terrific observations and yearning for science), TAPS (total group, alone, partners and small groups), Turn and Talk, Tutoring, and TV

Unit (advance) organizers and summaries

Values clarification, Venn diagramming, Video viewing (and production), Vignettes, Virtual field trips, Visuals, and Visualization techniques

Wait Time (or pause principle), WebQuest, Whiteboarding, Wikis, Word dissections, and Writing-to-learn exercises

X-word puzzles that encourage students to "Xplore and Xplain" science concepts

YRU still thinking teaching is just telling (or worse, maybe even Yelling?) and Yes/no cards

Zoom in on FUNdaMENTAL concepts to help students "see the forest for the trees" and Zoo mobiles and field trips that bring living creatures to schools and take students out to creatures

Web searches for specific instructional strategies will uncover a wealth of "how to use it" resources. In some instances, university-based centers are devoted to the promotion of one kind of instructional strategy (e.g., PBL, case study, etc.). Also, both general and science methods textbooks may devote entire chapters to specific methods. As a starting point for further reading, consider the following representative resource books:

Gregory, G. H., and E. Hammerman. 2008. *Differentiated instructional strategies for science, Grades K–8.* Thousand Oaks, CA: Corwin Press.

Herr, N. 2008. *The sourcebook for teaching science, grades 6–12: Strategies, activities, and instructional resources.* San Francisco, CA: Jossey-Bass.

Joyce, B., M. Weil, and E. Calhoun. 2008. *Models of teaching*. Boston, MA: Allyn and Bacon.

Lemov, D. 2010. *Teach like a champion: 49 techniques that put students on the path to college*. San Francisco, CA: Jossey-Bass.

Marzano, R. J., D. J. Pickering, and J. E. Pollock. 2001. *Classroom instruction that works: Research-based strategies for increasing student achievement*. Alexandria, VA: Association for Supervision and Curriculum Development.

Silberman, M. 1996. *Active learning: 101 strategies to teach any subject*. Boston: Allyn and Bacon. See also the follow-up book: Silberman, M. 2005. *Teaching actively: Eight steps and 32 strategies to spark learning in any classroom*. Boston: Allyn and Bacon.

Table A

		Low Minds-on	High Minds-on
Hands-on	**High**	Ill-designed and/or poorly guided instruction (given preparation/ability of learners) Illusion of inquiry[1]	Inquiry-oriented, guided discovery and experientially based learning (may morph into independent research)[1] High-quality teacher-scaffolded instruction (supports student construction of meaning)
	Low	Low-quality teacher-centered direct instruction (*Unprincipled presentations*[2])	High-quality teacher-directed, cognitively interactive instruction (*Principled presentations*[2])

Minds-on

1. See, for example, Singer, Hilton, and Schweingruber (2006).
2. Mayer (in Tobias and Duffy [2009]) presents a 2 × 2 table similar to the above that uses the phrase *behavioral activity* rather than *hands-on* and links the phrase *pure discovery* (in school contexts) with ineffective instruction. John Dewey (1938) made a similar distinction:

   Overemphasis upon activity as an end, instead of *intelligent* activity, leads to identification of freedom with immediate execution of impulses and desires (p. 69) … guidance given by the teacher to the exercise of the pupils' intelligence is an aid to freedom, not a restriction upon it … It is impossible to understand why a suggestion from one who has a larger experience and wider horizon should not be at least as valid as a suggestion arising from some more or less accidental source." (p. 71)

An Integrated, "Intelligent" Curriculum-Instruction-Assessment (CIA) System

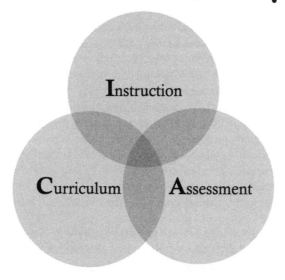

Curriculum: Standards to Benchmarks to Guides to Syllabi to Unit Plans *correlate* with …

Instruction: Intentionally designed sequences of learning activities with *differentiated* strategies and metacognitive and technological aids that scaffold student learning within their zone of proximal development and that are *aligned* with the …

Assessment: Diagnostic, formative, and summative evaluation of the ongoing development of students' cognitive, declarative knowledge ("know what"), process and/ or psychomotor skills (and/or procedural "know how"), and affective attitudes and habits of mind. It includes both formal and informal probes that use different types of metrics, are *embedded* in the curriculum and instruction, and provide "just in time" *feedback* to students and teachers to allow them to modify their subsequent learning and teaching actions. A complete assessment system provides relevant student-, teacher-, program-, and district-level data to administrators, parents, and other stakeholders for purposes of accountability and continuous improvement (NRC 2001b; NRC 2001c; NRC 2006).

Combined, these three mutually interdependent elements are the Central Intelligence Agency (CIA) of effective schools and classrooms. Research-informed, smart CIA is characterized by its creative balancing of subject knowledge–centered, learner-centered, assessment-centered, and community-centered perspectives (Donovan and Bransford 2005) and its focus on meaningful learning that is retained and can be transferred and creatively applied to new contexts (rather than inflexible, short-term rote learning). The whole of an integrated CIA system is greater than the sum of its parts due to the synergistic relationships between its three components. In this context, it is important to keep in mind the key characteristics of a "system" (NRC 2006, p. 27); namely, systems are

> organized around a specific goal; … composed of subsystems, or parts, that each serve their own purpose but also interact with other parts in ways that help the larger system function as intended; The subsystems that comprise the whole must work well both independently and together for the system to function as intended; The parts working together can perform functions that individual components cannot perform on their own; and A missing or poorly operating part may cause a system to function poorly, or not at all … they need to have well-developed feedback loops to keep the system from over- or underreacting to changes in a single elements. Feedback loops occur whenever part of an output of some system is connected back to one of its inputs.

Though published standards, guides, and textbooks are potentially wonderful teaching resources, effective teachers are not passive consumers of teacher-proof curricula and assessments or mere instructional technicians. Instead, they are critical reviewers and creative producers of CIA that they adapt (versus merely adopt) for the specific students and the learning communities that they serve. The 5E Teaching Cycle, developed by the Biological Curriculum Study Group, is a powerful pedagogical model that incorporates an integrated view of CIA (Bybee et al. 2006; O'Brien 2010, Appendix B). Balanced CIA can be represented by a variety of visuals, including a Venn diagram (on p. 267); an equilateral triangle framed by three points connected by a rubber band (Gross 1998; if only one of the three anchor points is moved, the triangle becomes distorted); and M.C. Escher works such as *Reptiles* and *Drawing Hand* that can represent the need for integrative, iterative cycles of CIA (*www.mcescher.*

com). In all cases, it is important to note that CIA is a subsystem of a larger system that includes pre- and inservice professional development and local school, district, state, and federal resources and regulations. Reforming and revitalizing science education will require systemwide considerations, but individual, dedicated science teachers working on the "front lines" of education can begin to turn the tide of the battle for "science for all Americans" as soon as the next class they teach or the next collegial conversation and collaboration they initiate.

Big Ideas in Science
A Comparison Across Science Standards Documents

Major reform documents in science education are in agreement that the following big ideas in science (along with key foundational theories) should serve as the conceptual lenses and framework (or "forest") to help contextualize the larger number of concepts ("trees") that make up a research-informed learning progression across grades K–12. Effective, integrated science curriculum-instruction-assessment (CIA) systems should be designed to help learners "uncover" misconceptions, "recover" valid prior knowledge, and "discover" new science concepts—not merely "cover" the pages of the textbooks. If the following "big ideas" are systematically developed and routinely referenced, lasting and transferable learning is more likely to occur. Less can become more when it comes to the content of the curriculum and learning outcomes.

AAAS Benchmarks for Science Literacy	National Science Education Standards	Foundation for Science Education
Four Common Themes (ch. 11)	Five Unifying Concepts and Processes	Seven Cross-Cutting Scientific Concepts
Systems*: parts ←→ whole interactions, input and output, controls, and positive and negative feedback, emergent properties, and subsystems	Systems, order, and organization	Systems and system models
Models: to study systems that are at different scales of time, size dimensions, etc., and the pros and cons of physical, conceptual, and mathematical models	Evidence, models, and explanation	See above box and Cause and effect: mechanisms and predictions
Constancy and change: stability, conservation, equilibrium, steady state and symmetry; rates, patterns and scale of change and graphical analysis, evolution, indeterminism, etc.	Change, constancy, and measurement	Stability and change -------------------------------- Energy and matter: flows, cycles, and conservation
See two boxes above	Evolution and equilibrium	Patterns, similarity, and diversity (e.g., classification and mathematical representations)
Scale: scale effects and mathematics, statistics, exponential notation, and powers of 10		Scale, proportion, and quantity
	Form and function	Form and function

* Note: Terminology in common across the three documents is italicized in the above table.

Sources

American Association for the Advancement of Science (AAAS). 1993. *Benchmarks for science literacy.* New York: Oxford University Press. *www.project2061.org/publications/default.htm.*

National Research Council (NRC). 1996. *National science education standards.* Washington, DC: National Academies Press. *www.nap.edu/catalog.php?record_id=4962.*

National Research Council (NRC). 2010. *A framework for science education: Preliminary public draft.* Washington, DC: National Academies Press. The final version of this report will be available in 2011 from the Board on Science Education, Division of Behavioral and Social Sciences and Education. *http://www7.nationalacademies.org/bose.*

Appendix D

Science Content Topics

BIOLOGICAL SCIENCES

Botany

6.   Resurrection Plant: Making Science Come Alive!

9.   Burdock and Velcro: Mother Nature Knows Best

Cell Theory

7.   Glue Mini-Monster: Wanted Dead or Alive?

10.   Osmosis and "Naked" Eggs: The Environment Matters

Characteristics of Life

6.   Resurrection Plant: Making Science Come Alive!

7.   Glue Mini-Monster: Wanted Dead or Alive?

8.   Water "Stick-to-It-Ness": A Penny for Your Thoughts (Extension #3)

Chromatography, Plant Pigments, and pH

11.   5 E(z) Yet pHenomenal Steps to Demystifying Magic Color-Changing Markers

Evolution and Biological Adaptations

6.   Resurrection Plant: Making Science Come Alive!

9.   Burdock and Velcro: Mother Nature Knows Best

Chromatography, Plant Pigments, and pH

11. 5 E(z) Yet pHenomenal Steps to Demystifying Magic Color-Changing Markers

Earth Science/Geology and Paleontology

 5. Ambiguous Text: Meaning-Making in Reading and Science (Extension #1: Fossil Footprints)

12. 5 E(z) Steps Back Into "Deep" Time: Visualizing the Geobiological Timescale

Light and Optics

 1. Science and Art: Dueling Disciplines or Dynamic Duo? (Optional Teacher Demonstration)

Physical and Chemical Changes

11. 5 E(z) Yet pHenomenal Steps to Demystifying Magic Color-Changing Markers

Process and Reasoning Skills for Teaching-Learning Science

Measurement, Magnitude (Scale), and Mathematical Skills

10. Osmosis and "Naked" Eggs: The Environment Matters

12. 5 E(z) Steps Back Into "Deep" Time: Visualizing the Geobiological Timescale

13. 5 E(z) Steps to Earth-Moon Scaling: Measurements and Magnitudes Matter

Nature of Science (NOS)*

 1. Science and Art: Dueling Disciplines or Dynamic Duo?

 2. Acronyms and Acrostics Articulate Attributes of Science (and Science Teaching)

* The NOS as "a way of knowing," is a cross-disciplinary theme or "big idea" that runs through all the activities.

Pseudoscience and/or Critical Thinking

Reading to Learn Science and Learning to Read: Science and English Language Arts Skills

Research Cited

Abd-El-Khalik, F., R. L. Bell, and N. G. Lederman. 1998. The nature of science and instructional practice: Making the unnatural natural. *Science Education* 82 (4): 417–436.

Aicken, F. 1991. *The nature of science.* 2nd ed. Portsmouth, NH: Heinemann.

Allchin, D. 2004. Error and the nature of science. ActionBioscience.org. *www. actionbioscience.org/education/allchin2.html.*

American Association for the Advancement of Science (AAAS). 1993. *Benchmarks for science literacy.* New York: Oxford University Press. *http://project2061.aaas. org.*

American Psychological Association (APA). 1997. *Learner-centered psychological principles: A framework for school redesign and reform.* Washington, DC: APA Center for Psychology in Schools and Education. *www.apa.org/ed/cpse/LCPP. pdf.*

Armstrong, T. 2000. *Multiple intelligences in the classroom.* 2nd ed. Alexandria, VA: Association for Supervision and Curriculum Development.

Baddock, M., and R. Bucat. 2008. Effectiveness of a classroom chemistry demonstration using the cognitive conflict strategy. *International Journal of Science Education* 30 (8): 1115–1128.

Banilower, E. R., S. E. Boyd, J. D. Pasley, and I. R. Weiss. 2006. *Lessons from a decade of mathematics and science reform: The local systemic change through teacher enhancement initiative.* Chapel Hill, NC: Horizon Research. *www.pdmathsci. net/findings/report/32.*

Banilower, E., K. Cohen, J. Pasley, and I. Weiss. 2008. *Effective science instruction: What does research tell us?* Portsmouth, NH: RMC Research Corporation, Center on Instruction. *www.centeroninstruction.com.*

Battino, R. 1979. Participatory lecture demonstrations. *Journal of Chemical Education* 56 (1): 39–41.

Becker, B. 1993. *20 demonstrations guaranteed to knock your socks off!* Batavia, IL: Flinn Scientific. See also Volume 2.

Begoray, D. L., and A. Stinner. 2005. Representing science through historical drama: Lord Kelvin and the age of the Earth debate. *Science & Education* 14 (3–5): 457–471. *www.springerlink.com/content/x777w6w788986637.*

Bell, R. L. 2008. *Teaching the nature of science through process skills: Activities for grades 3–8.* Boston, MA: Pearson/Allyn and Bacon.

Bilash, B. 1997. *A demo a day for physical science.* Batavia, IL: Flinn Scientific.

Bilash, B., and M. Shields. 2001. *A demo a day—A year of biological demonstrations.* Batavia, IL: Flinn Scientific.

Research Cited

Bilash, B., G. R. Gross, and J. K. Koob. 2006. *A demo a day—A year of chemical demonstrations.* 2nd ed. Batavia, IL: Flinn Scientific. See also Volume 2.

Bransford, J. D., A. L. Brown, and R. R. Cocking, eds. 1999a. *How people learn: Brain, mind, experience and school.* Washington, DC: National Academies Press.

Bransford, J. D., A. L. Brown, and R. R. Cocking, eds. 1999b. *How people learn: Bridging research and practice.* Washington, DC: National Academies Press.

Brooks, J. G., and M. G. Brooks. 1999. *In search of understanding: The case for constructivist classrooms.* Alexandria, VA: Association for Supervision and Curriculum Development.

Buckman, R. 2003. *Human wildlife: The life that lives on us.* Baltimore, MD: Johns Hopkins University Press.

Bybee, R. W., ed. 2002. *Learning science and the science of learning.* Arlington, VA: NSTA Press.

Bybee, R. W., J. A. Taylor, A. Gardner, P. Van Scotter, J. Carlson Powell, A. Westbrook, and N. Landes. 2006. *BSCS 5E Instructional Model: Origins, effectiveness and applications.* Colorado Springs, CO: BSCS. *www.bscs.org/pdf/5EFull Report.pdf* (65 pages). *http://bscs.org/pdf/bscs5eexecsummary.pdf* (19 pages).

Bybee, R. W., J. C. Carlson Powell, and L. W. Trowbridge. 2008. *Teaching secondary school science: Strategies for developing scientific literacy.* 9th ed. Upper Saddle River, NJ: Merrill Prentice Hall.

Calaprice, A., ed. 2011. *The ultimate quotable Einstein.* Princeton, NJ: Princeton University Press.

Camp, C. W., and J. J. Clement. 1994. *Preconceptions in mechanics: Lessons dealing with students' conceptual difficulties.* Dubuque, IA: Kendall/Hunt.

Carter, G. S., and R. D. Simpson. 1978. Science and reading: A basic duo. *The Science Teacher* 45 (3): 20.

Chamberlin, T. C. 1965 (reprint from 1890). The method of multiple working hypotheses. *Science* 148: 754–759.

Chiappetta, E. L., and T. R. Koballa Jr. 2010. *Science instruction in the middle and secondary school: Developing fundamental knowledge and skills for teachers.* 7th ed. Boston: Allyn and Bacon.

Chin, C., and J. Osborne. 2008. Students' questions: A potential resource for teaching and learning science. *Studies in Science Education* 44 (1): 1–39.

Chinn, C. A., and W. F. Brewer. 1993. The role of anomalous data in knowledge acquisition: A theoretical framework and implications for science instruction. *Review of Educational Research* 63 (1): 1–49.

Chinn, C. A., and W. F. Brewer. 1998. An empirical test of a taxonomy of responses to anomalous data in science. *Journal of Research in Science Teaching* 35 (6): 623–654.

Clark, R. L., M. P. Clough, and C. A. Berg. 2000. Modifying cookbook labs: A different way of teaching a standard laboratory engages students and promotes understanding. *The Science Teacher* 67 (7): 40–43.

Clough, M. P. 2002. Using the laboratory to enhance student learning. In *Learning science and the science of learning*, ed. R. W. Bybee, 85–94. Arlington, VA: NSTA Press.

Clough, M. P. 2004. The nature of science: Understanding how the game is played. In *The game of science education,* ed. J. Weld, 198–227. Boston, MA: Pearson/ Allyn and Bacon.

Clough, M. P., and R. L. Clark. 1994a. Cookbooks and constructivism: A better approach to laboratory activities. *The Science Teacher* 61 (2): 34–37.

Clough, M. P., and R. L. Clark. 1994b. Creative constructivism: Challenge your students with authentic science experience. *The Science Teacher* 61 (7): 46–49.

Coalition of Essential Schools Northwest. n.d. Critical friends groups. *www. cesnorthwest.org/cfg.php.*

Cochran, K. F. 1997. Pedagogical content knowledge: Teacher's integration of subject matter, pedagogy, students, and learning environments. Brief. *Research Matters to the Science Teacher.* No. 9702. National Association in Research in Science Teaching. *www.narst.org/publications/research/pck.htm.*

Cocking, R. R., J. P. Mestre, and A. L. Brown, eds. 2000. New developments in the science of learning: Using research to help students learn science and mathematics. *Journal of Applied Developmental Psychology, Special Issue* 21 (1): 1–135.

Colburn, A. 2004. Focusing labs on the nature of science: Laboratories can be structured to help students better understand the nature of science. *The Science Teacher* 71 (9): 32–35.

Council of Chief State School Officers (CCSSO). 2010. *Model core teaching standards: A resource for state dialogue.* Draft revision of the 1992 standards developed by the Interstate New Teacher Assessment and Support Consortium (INTASC). Washington, DC: CCSSO. *www.ccsso.org/Resources/Publications/Model_Core_ Teaching_Standards.html.*

Council of State Science Supervisors (CSSS). Science and safety guides (free downloads). *www.csss-science.org/safety.shtml.*

Cromer, A. 1993. *Uncommon sense: The heretical nature of science.* New York: Oxford University Press.

Cunningham, J., and N. Herr. 2002. *Hands-on chemistry activities with real-life applications: Easy-to-use labs and demonstrations for grades 8–12.* San Francisco: Jossey-Bass.

DeBoer, G. E. 1991. *A history of ideas in science education: Implications for practice.* New York: Teachers College Press.

Dewey, J. (1910) 1991. *How we think.* Amherst, NY: Prometheus Books. *www. brocku.ca/MeadProject/Dewey/Dewey_1910a/Dewey_1910_a.html.*

Dewey, J. 1913. *Interest and effort in education.* Boston, MA: Houghton Mifflin Co. *www.archive.org/details/interestandeffor00deweuoft.*

Dewey, J. (1916) 1924. *Democracy and education: An introduction to the philosophy of education.* New York: Macmillan. *www.ilt.columbia.edu/Publications/dewey.html.*

Dewey, J. 1938. *Experience and education.* New York: Macmillan Publishing.

Dobell, C. 1958. *Antony van Leeuwenhoek and his "Little Animals": Being some account of the father of protozoology and bacteriology and his multifarious discoveries in these*

Research Cited

disciplines. With an introduction by C. B. van Neil. New York: Russell and Russell.

Donovan, M. S., and J. D. Bransford, eds. 2005. *How students learn: Science in the classroom.* Washington, DC: National Academies Press.

Doran, R., F. Chan, P. Tamir, and C. Lenhardt. 2002. *Science educator's guide to laboratory assessment.* Arlington, VA: NSTA Press.

Driver, R., E. Guesne, and A. Tiberghein, eds. 1985. *Children's ideas in science.* Milton Keynes, England: Open University Press.

Driver, R., A. Squires, P. Rushworth, and V. Wood-Robinson. 1994. *Making sense of secondary science: Research into children's ideas.* London: Routledge.

Dubin, J. 2009. Growing together: American teachers embrace the Japanese art of lesson study. *American Educator* 33 (3): 30–34. *www.aft.org/newspubs/periodicals/ae/fall2009/index.cfm.*

DuFour, R., and R. Eaker. 1998. *Professional learning communities at work: Best practices for enhancing student achievement.* Bloomington, IN: National Educational Service and Alexandria, VA: Association for Supervision and Curriculum Development.

Duit, R. 2009. *Bibliography of students' and teachers' conceptions and science education:* Kiel, Germany: Institute for Science Education at the University of Kiel. *www.ipn.uni-kiel.de/aktuell/stcse/stcse.html.*

Ehrlich, R. 1990. *Turning the world inside out and 174 other simple physics demonstrations.* Princeton, NJ: Princeton University Press.

Ehrlich, R. 1997. *Why toast lands jelly-side down: Zen and the art of physics demonstrations.* Princeton, NJ: Princeton University Press.

Enger, S. K., and R. E. Yager. 2001. *Assessing student understanding in science: A standards-based K–12 handbook.* Thousand Oaks, CA: Corwin Press.

Fensham, P., R. Gunstone, and R. White, eds. 1994. *The content of science: A constructivist approach to its teaching and learning.* London: Falmer Press.

Fisher, K. M., and J. L. Lipson. 1986. Twenty questions about student errors. *Journal of Research in Science Teaching* 23 (9): 783–803.

Flinn Scientific, Inc. Safety resources: *www.flinnsci.com/Sections/Safety/safety.asp.*

Freier, G. D., and F. J. Anderson. 1981. *A demonstration handbook for physics.* 2nd ed. College Park, MD: American Association of Physics Teachers. *www.aapt.org/publications.*

Gallagher, J. J. 2007. *Teaching science for understanding: A practical guide for middle and high school teachers.* Upper Saddle River, NJ: Merrill Prentice Hall.

Gardner, H. 1999. *The disciplined mind: What all students should understand.* New York: Simon and Schuster.

Garet, M. S., A. C. Porter, L. Desimone, B. F. Birman, and K. S. Yoon. 2001. What makes professional development effective? Results for a national sample of teachers. *American Educational Research Journal* 38 (4): 915–945.

Gilbert, S. W., and S. Watt Ireton. 2003. *Understanding models in Earth and space science.* Arlington, VA: NSTA Press.

Goleman, D. 1995. *Emotional intelligence: Why it can matter more than IQ.* New York: Bantam Books.

Good, R. G., J. D. Novak, and J. H. Wandersee, eds. 1990. Perspectives on concept mapping. *Journal of Research in Science Teaching, Special Issue* 27 (10). *www.narst.org.*

Grant, J. 2006. *Discarded science: Ideas that seemed good at the time.* Wisley, Surrey, UK: Facts, Figures and Fun.

Gross, G. R., M. A. Holzer, and E. A. Colangelo. 2001. *A demo a day—A year of Earth science demonstrations.* Batavia, IL: Flinn Scientific.

Gross, S. J. 1998. *Staying centered: Curriculum leadership in a turbulent era.* Alexandria, VA: Association for Supervision and Curriculum Development.

Grun, B., and E. Simson. 2005. *The timetables of history: A horizontal linkage of people and events.* 4th ed. New York: Simon and Schuster/Touchstone Books.

Hackney, M. W., and J. H. Wandersee. 2002. *The power of analogy: Teaching biology with relevant classroom-tested activities.* Reston, VA: National Association of Biology Teachers.

Hagevvik, R., W. Veal, E. M. Brownstein, E. Allen, C. Ezrailson, and J. Shane. 2010. Pedagogical content knowledge and the 2003 science teacher preparation standards for NCATE accreditation or state approval. *Journal of Science Teacher Education* 21 (1): 7–12.

Harms, N. C., and R. E. Yager, eds. 1981. *What research says to the science teacher (project synthesis).* Vol. 3. Washington, DC: NSTA Press.

Harris Freedman, R. L. 1999. *Science and writing connections.* White Plains, NY: Dale Seymour Publications.

Harrison, A. G., and R. K. Coll, eds. 2008. *Using analogies in middle and secondary science classrooms: The FAR guide—An interesting way to teach with analogies.* Thousand Oaks, CA: Corwin Press.

Harvard-Smithsonian Center for Astrophysics, Science Education Department. MOSART: Misconception Oriented Standards-based Assessment Resource: *www.cfa.harvard.edu/sed/projects/mosart.html.* See also: *Minds of Own* and *A Private Universe.*

Hilton, M. (rapporteur). 2010. *Exploring the intersection of science education and 21st century skills: A workshop summary.* Washington, DC: National Academies Press. *http://books.nap.edu/catalog.php?record_id=12771.*

Hoagland, M., and B. Dodson. 1998. *The way life works: The science lover's illustrated guide to how life grows, develops, reproduces, and gets along.* New York: Three Rivers Press/Random House.

Hofstein, A., and V. N. Lunetta. 2003. The laboratory in science education: Foundation for the Twenty-first century. *Science Education* 88 (1): 28–54.

Hord, S., and W. Summers. 2008. *Leading professional learning communities: Voices from research and practice.* Thousand Oaks, CA: Corwin.

Ingersoll, R. M. 2003. Turnover and shortages among science and mathematics teachers in the United States, In *Science teacher retention: Mentoring and renewal,* ed. J. Rhoton and P. Bowers, 1–12. Arlington, VA: NSTA Press.

Ingram, M. 2003. *Bottle biology: Exploring the world through soda bottles and other recyclable materials.* 2nd ed. Dubuque, IA: Kendall Hunt.

Research Cited

Jewett, J. W., Jr. 1994. *Physics begins with an M ... Mysteries, magic and myth.* Boston, MA: Allyn and Bacon.

Jewett, J. W., Jr. 1996. *Physics begins with another M ... .* Boston, MA: Allyn and Bacon.

Johnson, S. 2008. *The invention of air: A story of science, faith, revolution and the birth of America.* New York Riverhead Books/Penguin Group.

Kardos, T. 2003. *Easy science demos and labs—Life science.* 2nd ed. Portland, ME: J. Weston Walch Education. See also other titles in series: *Chemistry, Earth Science,* and *Physics.*

Keeley, P., F. Eberle, and L. Farrin. 2005. *Uncovering student ideas in science, Volume 1: 25 formative assessment probes.* Arlington, VA: NSTA Press.

Keeley, P., F. Eberle, and J. Tugel. 2007. *Uncovering student ideas in science, Volume 2: 25 more formative assessment probes.* Arlington, VA: NSTA Press.

Kind, V. 2004. *Beyond appearance: Students' misconceptions about basic chemical ideas.* 2nd ed. *www.rsc.org/education/teachers/learnnet/pdf/LearnNet/rsc/miscon.pdf.*

Klentschy, M. P. 2010. *Using science notebooks in middle school.* Arlington, VA: NSTA Press.

Kwan, T., and J. Texley. 2003. *Exploring safely: A guide for middle school teachers.* Arlington, VA: NSTA Press.

Lawson, A. E., ed. 1993. The role of analogy in science and science teaching. *Journal of Research in Science Teaching, Special Issue* 30 (10). *www.narst.org.*

Lawson, A. E. 2010. *Teaching inquiry science in middle and secondary schools.* Thousand Oaks, CA: Sage Publications.

Lederman, N. G. 1992. Students' and teachers' conceptions of the nature of science: A review of the research. *Journal of Research in Science Teaching* 29 (4): 331–359.

Lederman, N. G. 1999. Teachers' understanding of the NOS and classroom practice: Factors that facilitate or impede the relationship. *Journal of Research in Science Teaching* 36 (8): 916–929.

Lederman, N. G., and M. L. Neiss. 1997. The nature of science: Naturally? *School Science and Mathematics* 97 (1): 1–2.

Levitin, D. J. 2006. *This is your brain on music: The science of a human obsession.* New York: Plume/Penguin Books. See also the author's Laboratory for Music Perception, Cognition, and Expertise at McGill University: *www.psych.mcgill. ca/labs/levitin.*

Lewis, C. C., and I. Tsuchida. 1998. A lesson is like a swiftly flowing river: How research lessons improve Japanese education. *American Educator* 22 (4): 12–17, 50–52. *www.aft.org/newspubs/periodicals/ae/winter1998/index.cfm.*

Liu, X. 2010. *Essential of science classroom assessment.* Thousand Oaks, CA: SAGE Publications.

Lortie, D. C. 1975. *Schoolteacher: A sociological study.* Chicago: University of Chicago Press.

Loucks-Horsley, S., P. W. Hewson, N. Love, and K. E. Stiles. 1998. *Designing professional development for teachers of science and mathematics.* National Institute for Science Education. Thousand Oakes, CA: Corwin Press.

Louv, R. 2008. *Last child in the woods: Saving our children from nature-deficit disorder.* Chapel Hill, NC: Algonquin Books.

Lunetta, V. N., A. Hofstein, and M. Clough. 2007. Learning and teaching in the school science laboratory. In *Handbook of research in science education,* ed. N. Lederman and S. Abell, 393–441. Mahwah, NJ: Lawrence Erlbaum.

Mathewson, J. H. 1999. Visual-spatial thinking: An aspect of science overlooked by educators. *Science Education* 83: 33–54. *www.citeulike.org/user/oritpa/article/2320205.*

Mayer, R. E. 2009a. *Multimedia learning.* 2nd ed. New York: Cambridge University Press.

Mayer, R. E. 2009b. Constructivism as a theory of learning versus constructivism as a prescription for instruction. In *Constructivist instruction: Success or failure?* ed. S. Tobias and T. M. Duffy, 184–200. New York: Routledge

McComas, W. F. 1996. Myths of science: Reexamining what we know about the nature of science. *School Science and Mathematics* 96 (1): 10–16.

McComas, W. F., ed. 1998. *The nature of science in science education: Rationales and strategies.* Dordecht, The Netherlands: Kluwer Academic Publishers.

McComas, W. F., ed. 2004. The history and nature of science. *The Science Teacher, Special Issue* 71 (9).

McCombs, B. L., and J. S. Whisler. 1997. *The learner-centered classroom and school: Strategies for increasing student motivation and achievement.* San Francisco, CA: Jossey-Bass.

Meaningful Learning Research Group. Misconceptions conference proceedings. *http://www2.ucsc.edu/mlrg/mlrgarticles.html.*

Michael, J. A., and H. I. Modell. 2003. *Active learning in secondary and college science classrooms: A working model for helping the learner to learn.* Mahwah, NY: Lawrence Erlbaum Associates.

Michaels, S., A. W. Shouse, and H. A. Schweingruber. 2008. *Ready, set, science! Putting research to work in K–8 science classrooms.* Washington, DC: National Academies Press. *www.nap.edu/catalog.php?record_id=11882.*

Mintzes, J. J., J. H. Wandersee, and J. D. Novak, eds. 1998. *Teaching science for understanding: A human constructivist view.* New York: Academic Press.

Mintzes, J. J., J. Wandersee, and J. D. Novak, eds. 2000. *Assessing science understanding: A human constructivist view.* San Diego, CA: Academic Press.

Morowitz, H. J. 2002. *The emergence of everything: How the world became complex.* New York: Oxford University Press.

Mundry, S., and K. Stiles, eds. 2009. *Professional learning communities for science teaching: Lessons from research and practice.* Arlington, VA: NSTA Press.

National Academy of Sciences (NAS). 1998. *Teaching about evolution and the nature of science.* Washington, DC: National Academies Press.

National Academy of Sciences (NAS). 2007. *Rising above the gathering storm: Energizing and employing America for a brighter economic future.* Washington, DC: National Academies Press. *www.nap.edu/catalog.php?record_id=11463.*

Research Cited

National Board for Professional Teaching Standards (NBPTS). 2003a. *NBPTS Science/Adolescent and young adulthood standards for teachers of students ages 14–18+*. 2nd ed. *www.nbpts.org/the_standards/standards_by_cert*.

National Board for Professional Teaching Standards (NBPTS). 2003b. *NBPTS Science/Adolescent and young adulthood standards for teachers of students ages 11–15*. 2nd ed. *www.nbpts.org/the_standards/standards_by_cert*.

National Commission on Excellence in Education (NCEE). 1983. *A nation at risk: The imperative for education reform* (Stock No. 065-000-001772). Washington, DC: U.S. Government Printing Office. *www.ed.gov/pubs/NatAtRisk/index.html*.

National Commission on Mathematics and Science Teaching (NCMST) for the 21st Century. 2000. *Before it's too late*. Washington, DC: Department of Education. *www.ed.gov/inits/Math/glenn/report.pdf*.

National Commission on Teaching and America's Future (NCTAF). 1996. *What matters most: Teaching for America's future*. New York: Teachers College, Columbia University. *www.nctaf.org/documents/WhatMattersMost.pdf*.

National Commission on Teaching and America's Future (NCTAF). 1997. *Doing what matters most: Investing in quality teaching*. New York: Teachers College, Columbia University. *www.nctaf.org/documents/DoingWhatMattersMost.pdf*.

National Research Council (NRC). 1996. *National science education standards*. Washington, DC: National Academies Press. *www.nap.edu/catalog.php?record_id=4962*.

National Research Council (NRC). 2000. *Inquiry and the national science education standards: A guide for teaching and learning*. Washington, DC: National Academies Press. *www.nap.edu/catalog.php?record_id=9596*.

National Research Council (NRC). 2001a. *Educating teachers of science, mathematics, and technology: New practices for the new millennium*. Washington, DC: National Academies Press. *www.nap.edu/catalog.php?record_id=9832*.

National Research Council (NRC). 2001b. *Classroom assessment and the national science education standards*. Washington, DC: National Academies Press. *www.nap.edu/catalog.php?record_id=9847*.

National Research Council (NRC). 2001c. *Knowing what students know: The science and design of educational assessment*. Washington, DC: National Academies Press. *www.nap.edu/catalog.php?record_id=10019*.

National Research Council (NRC). 2006. *Systems for state science assessment*. Washington, DC: National Academies Press. *www.nap.edu/catalog.php?record_id=11312*.

National Research Council (NRC). 2007. *Taking science to school: Learning and teaching science in grades K–8*. Washington, DC: National Academies Press. *www.nap.edu/catalog.php?record_id=11625*.

National Research Council (NRC). 2010a. *A framework for science education: Preliminary public draft*. Washington, DC: National Academies Press. The final version of this report will be available in 2011 from the Board on Science Education, Division of Behavioral and Social Sciences and Education. *http://www7.nationalacademies.org/bose*.

National Research Council (NRC). 2010b. *Preparing teachers: Building evidence for sound policy.* Washington, DC: National Academies Press. *www.nap.edu/ catalog.php?record_id=12882.*

National Science Board (NSB). 2006. *America's pressing challenge—Building a stronger foundation.* Washington, DC: National Science Foundation. NSB 06-02. *www. nsf.gov/statistics/nsb0602.*

National Science Teachers Association (NSTA). 2000. *The nature of science.* NSTA Position Paper adopted in July 2000. Arlington, VA: NSTA. *www.nsta.org/ about/positions/natureofscience.aspx.*

National Science Teachers Association (NSTA). 2003/Revised 2010. *Standards for science teacher preparation.* Arlington, VA: NSTA. *www.nsta.org/preservice.*

National Science Teachers Association (NSTA). 2004a. *Science inquiry.* NSTA Position Paper adopted in October 2004. Arlington, VA: NSTA. *www.nsta. org/about/positions/inquiry.aspx.*

National Science Teachers Association (NSTA). 2004b. *Science teacher preparation.* NSTA Position Paper adopted in July 2004. Arlington, VA: NSTA. *www.nsta. org/about/positions/preparation.aspx.*

National Science Teachers Association (NSTA). 2006. *Professional development in science education.* NSTA Position Paper adopted in May 2006. Arlington, VA: NSTA. *www.nsta.org/about/positions/profdev.aspx.*

National Science Teachers Association (NSTA). 2007a. *Principles of professionalism for science educators.* NSTA Position Paper adopted in June 2007. Arlington, VA: NSTA. *www.nsta.org/about/positions/professionalism.aspx.*

National Science Teachers Association (NSTA). 2007b. *Induction programs for the support and development of beginning teachers of science.* NSTA Position Paper adopted in April 2007. Arlington, VA: NSTA. *www.nsta.org/about/positions/ induction.aspx.*

National Science Teachers Association (NSTA). 2007c. *The integral role of laboratory investigations in science instruction.* NSTA Position Paper adopted in February 2007. Arlington, VA: NSTA. *www.nsta.org/about/positions/laboratory.aspx.*

National Science Teachers Association (NSTA). 2008. *The role of E-learning in science education.* NSTA Position Paper adopted in September 2008. Arlington, VA: NSTA. *www.nsta.org/about/positions/e-learning.aspx.*

National Science Teachers Association (NSTA). 2010a. *The role of research on science teaching and learning.* NSTA Position Paper adopted in September 2010. Arlington, VA: NSTA. *www.nsta.org/about/positions/research.aspx.*

National Science Teachers Association (NSTA). 2010b. *Teaching science and technology in the context of societal and personal issues.* NSTA Position Paper adopted in November 2010. Arlington, VA: NSTA. *www.nsta.org/about/ positions/societalpersonalissues.aspx.*

National Staff Development Council (NSDC). 2001. NSDC's standards for staff development. *www.learningforward.org/standards/index.cfm.*

Newcombe, N. S., 2010. Picture this: Increasing math and science learning by improving spatial thinking. *American Educator* 34 (2): 29–35. *www.aft.org/ newspubs/periodicals/ae/summer2010/index.cfm.*

Research Cited

Nisbett, R. E. 2009. *Intelligence and how to get it: Why schools and cultures count.* New York: W. W. Norton.

Norton-Meier, L., B. Hand, L. Hockenberry, and K. Wise. 2008. *Questions, claims, and evidence: The important place of argument in children's science writing.* Portsmouth, NH: Heinemann.

O'Brien, T. 1991. The science and art of demonstrations. *Journal of Chemical Education* 68: 933–936.

O'Brien, T. 1993. Teaching fundamental aspects of science toys. *School Science and Mathematics* 93: 203–207.

O'Brien, T. 2000. A toilet paper timeline of evolution: A 5E cycle on the concept of scale. *American Biology Teacher* 62 (8): 578–582.

O'Brien, T. 2010. *Brain-powered science: Teaching and learning with discrepant events.* Arlington, VA: NSTA Press.

O'Brien, T. 2011. *More brain-powered science: Teaching and learning with discrepant events.* Arlington, VA: NSTA Press.

O'Brien, T., and D. Seager. 2000. 5 E(z) steps to teaching Earth-Moon scaling: An interdisciplinary mathematics/science/ technology mini-unit. *School Science & Mathematics* 100 (7): 390–395.

Olenick, R. P. 2008. Comprehensive conceptual curriculum for physics (C3P) project: Misconceptions and preconceptions in introductory physics. *http://phys.udallas.edu/C3P/Preconceptions.pdf.*

Operation Physics: Children's misconceptions about science. *www.amasci.com/miscon/opphys.html.*

Osborne, R., and P. Freyberg. 1985. *Learning in science: The implications of children's science.* London: Heinemann.

Palmer, P. J. 2007. *The courage to teach: Exploring the inner landscape of a teacher's life.* 10th ed. San Francisco, CA: Jossey-Bass.

Panek, R. 1998. *Seeing and believing: How the telescope opened our eyes and minds to the heavens.* New York: Penguin Books.

Partnership for 21st Century Skills. 2009. *Framework for 21st century learning. www.p21.org/index.php?option=com_content&task=view&id=254&Itemid=120.*

Posner, G. J., K.A. Strike, P. W. Hewson, and W. A. Gertzog. 1982. Accommodation of a scientific conception: Toward a theory of conceptual change. *Science Education* 66: 211–227.

President's Council of Advisors on Science and Technology (PCAST). 2010. *Report to the President, Prepare and inspire: K–12 Education in science, technology, engineering and math (STEM) for America's future.* Prepublication version. *www.whitehouse.gov/administration/eop/ostp/pcast/docsreports.*

Rhoton, J., and P. Bowers, eds. 2003. *Science teacher retention: Mentoring and renewal.* Arlington, VA: NSTA Press.

Rotherham, A. J., and D. T. Willingham. 2010. 21st-century skills: Not new, but a worthy challenge. *American Educator* 34 (1): 17–20. Original source: *Educational Leadership* 67 (1): 16–21. *www.aft.org/newspubs/periodicals/ae/spring2010/index.cfm.*

Rowe, M. B., ed. 1978. *What research says to the science teacher.* Vol. 1. Arlington, VA: NSTA Press.

Rutherford, J., and A. Ahlgren. 1991. *Science for all Americans.* 2nd ed. New York: Oxford University Press. *www.project2061.org/publications/sfaa/default.htm.*

Sae, A. S. 1996. *Chemical magic from the grocery store.* Dubuque, IA: Kendall/Hunt.

Sarquis, M., L. Hogue, S. Hershberger, J. Sarquis, and J. Williams. 2009. *Chemistry with charisma: 24 lessons that capture and keep attention in the classroom.* Vol. 1 Middletown, OH: Terrific Science Press. *www.terrrrificscience.org/charisma.*

Sarquis, M., L. Hogue, S. Hershberger, J. Sarquis, and J. Williams. 2010. *Chemistry with charisma: 28 more lessons that capture and keep attention in the classroom.* Vol. 2. Middletown, OH: Terrific Science Press. *www.terrrrificscience.org/charisma.*

Sarquis, M., J. P. Williams, and J. L. Sarquis. 1995. *Teaching chemistry with toys: Activities for grades K–9.* New York: McGraw-Hill/Terrific Science Press.

Schön, D. A. 1983. *The reflective practitioner: How professionals think in action.* New York: Basic Books.

Science Hobbyist: *Amateur Science: Science myths in K–6 textbooks and popular culture* (with extensive links to other sites): *www.amasci.com/miscon/miscon.html.*

Shulman, L. 1986. Those who understand: Knowledge growth in teaching. *Educational Researcher* 15 (2): 4–14.

Shulman, L. 1987. Knowledge and teaching: Foundations of the new reform. *Harvard Educational Review* 57 (1): 1–22.

Singer, S. R., M. L. Hilton, and H. A. Schweingruber. 2006. *America's lab report: Investigations in high school science.* Washington, DC: National Academies Press. Executive Summary: *www.nap.edu/catalog/11311.html.*

STEM Education Coalition. See especially links to Reports. *http://nstacommunities.org/stemedcoalition.*

Stepans, J. 1994. *Targeting students' science misconceptions: Physical science activities using the conceptual change model.* Riverview, FL: Idea Factory.

Stigler, J. W., and J. Hiebert. 1999. *The teaching gap: Best ideas from the world's teachers for improving education in the classroom.* New York: Free Press/Simon and Schuster.

Summerlin, L. R., and J. L. Ealy. 1988. *Chemical demonstrations: A sourcebook for teachers.* Vol. 1. 2nd ed. Washington, DC: American Chemical Society. See also Volume 2 (with C. L. Borgford).

Taber, K. 2002. *Chemical misconceptions: Prevention, diagnosis and cure.* Vol. 1 and 2. Cambridge, UK: Royal Society of Chemistry. New York: Springer.

Tauber, R. T., and C. Sargent Mester. 2007. *Acting lessons for teachers: Using performance skills in the classroom.* 2nd ed. Westport, CT: Praeger.

Taylor, B., J. Poth, and D. Portman. 1995. *Teaching physics with toys: Activities for grades K–9.* Terrific Science Press. New York: McGraw-Hill.

Texley, J., T. Kwan, and J. Summers. 2004. *Exploring safely: A guide for high school teachers.* Arlington, VA: NSTA Press.

Research Cited

Tobias, S., and A. Baffert. 2010. *Science teaching as a profession: Why it isn't. How it could be.* Arlington, VA: NSTA Press. See also project website: *www.science-teaching-as-a-profession.com.*

Tobias, S., and T. M. Duffy, eds. 2009. *Constructivist instruction: Success or failure?* New York: Routledge.

Treagust, D. F., R. Duit, and B. J. Fraser, eds. 1996. *Improving teaching and learning in science and mathematics.* New York: Teachers College Press.

Trefil, J. 2008. *Why science?* New York: Teachers College Press and Arlington, VA: NSTA Press.

Twain, M. (1880) 2004. *A tramp abroad.* Reprint, Whitefish, MT: Kessinger Publishing.

U.S. Department of Education, Office of Educational Technology. 2010. *National education technology plan 2010: Transforming American education: Learning powered by technology.* Washington, DC: U.S. Department of Education. *www.ed.gov/technology/netp-2010.*

U.S. Department of Health and Human Services. Household products database. *http://householdproducts.nlm.nih.gov.*

VanCleave, J. P. 1989. *Janice VanCleave's chemistry for every kid: 101 easy experiments that really work.* Hoboken, NJ: John Wiley and Sons.

VanCleave, J. P. 1990. *Janice VanCleave's biology for every kid: 101 easy experiments that really work.* Hoboken, NJ: John Wiley and Sons.

VanCleave, J. P. 1991a. *Janice VanCleave's Earth science for every kid: 101 easy experiments that really work.* Hoboken, NJ: John Wiley and Sons.

VanCleave, J. P. 1991b. *Janice VanCleave's physics for every kid: 101 easy experiments that really work.* Hoboken, NJ: John Wiley and Sons.

Volkman, M. J., and S. K. Abell. 2003. Rethinking laboratories. *The Science Teacher* 70 (6): 38–41.

White, R. T., and R. F. Gunstone. 1992. *Probing understanding.* London: Falmer.

Wolpert, L. 1992. *The unnatural nature of science.* Cambridge, MA: Harvard University Press.

Yager, R. E. 1983. The importance of terminology in teaching K–12 science. *Journal of Research in Science Teaching* 20 (6): 577.

Yager, R. E., ed. 2005. *Exemplary science: Best practices in professional development.* Arlington, VA: NSTA Press.

Youngson, R. 1998. *Scientific blunders: A brief history of how wrong scientists can sometimes be.* New York: Carroll and Graf Publishers.

NATIONAL SCIENCE TEACHERS ASSOCIATION

Index

Index

Index

Index

safety note for, 126
science concepts in, xxvii, 116
science education concepts in,
116–117
*What Research Says to the Science
Teacher (Project Synthesis),* 10
Whitman, Walt, 234

Why Science?, 10
*Why We Get Sick: The New Science of
Darwinian Medicine,* 210
Williams, G. C., 210

Y

Yeast, 102, 113

Z

Zone of proximal development
(ZPD), 78